Lotus Illustrated DICTIONARY of ELECTRONICS & TELECOMMUNICATION

Complied and Edited by:
Camic James

Lotus PRESS
4735/22, Prakash Deep Building
Ansari Road, Darya Ganj,
New Delhi-110002

Lotus Illustrated

DICTIONARY of ELECTRONICS & TELECOMMUNICATION

ISBN 81 89093 31 2

Published by:

Lotus Press : Publishers & Distributors
Unit No. 220, 2nd Floor, 4735/22,
Prakash Deep Building, Ansari Road, Darya Ganj,
New Delhi- 110002, Ph.: 41325510, 09811838000
• E-mail : lotuspress1984@gmail.com www.lotuspress.co.in

Printed & Published by : **Lotus Press Publisher & Distributors,** New Delhi-02

PREFACE

The dictionary is an exhaustive and comprehensive list of the terms pertaining to the subject of electronics and its allied subject viz.instrumentation, transducers, electrical and all other related terms which are being widely accepted as electronics related. This book is specifically developed for engineers and research scholars of this field, because it contains terms that are being widely used in their day-to-day curriculum. It provides all sort of information and contains more than 1500 entries.

This electronics dictionary consists of all that material, which is generally required by them in their day-to-day assignments. Rich and exhaustive definitions are provided of the terms .The terms are supplemented by the pictorial information. The graphics supplemented are the best thing that is being put in this dictionary and are self-explaining and in a way provide alternate way of learning and grasping the subject more strongly. This dictionary of electronics contains terms of all fields attached to electronics, which are related, to electronics and not just of plane electronics. This dictionary also contains the terms, which are generally not found in other dictionary of the same stature. All in this entire dictionary will prove to be a best tool to learn the subject in great depth. A stress is put on including the various definitions, which are common to different topics and are dealt according to their relative topics. All the possible definitions are provided in the simplest manner which readers find very easy to adopt.

This dictionary is developed with great effort and lot of hard work. We hope that we will be successful in increasing your knowledge by leaps and bounds. A response about the book is widely anticipated. Wish you a very joyous ride to the electronics subject.

■ **µbar**

a unit of pressure used in acoustics.

■ **µPa (micro Pascal)**

a unit of pressure used in acoustics.

■ **2 quadrant analog multiplier**

a multiplier where the control input must be a positive voltage.

■ **4 quadrant analog multiplier**

a multiplier where both inputs (signal and control) may be positive or negative.

■ **555 timer**

an integrated circuit that can be used in one of two ways. As a monostable time changes from one state to another for a set period of time and then is required to be reset .As an astable timer it continually changes from one state to another turning on and off.

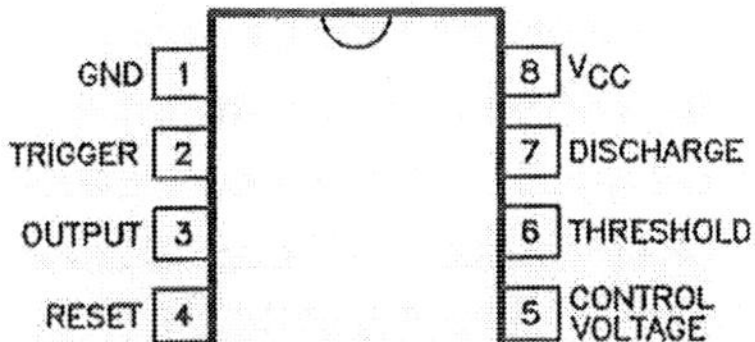

555

■ **a**

abbreviation for "ampere" a unit of electrical current.

■ **A/D (ANALOG-TO-DIGITAL) CONVERTER**

an electronic device, often an integrated circuit, that converts an analog voltage to a digital value. All digital instruments use an A/D converter to convert the input signal into digital information.

■ **A/D**

abbreviation for analog-to-digital.

■ **absolute measuring system**

the measuring value is determined by reading information from a scale, without counting. The measuring value is immediately available after switch-on.

■ **absorbance**

the ability of a medium to absorb radiation depending on temperature and wavelength. Expressed as the negative common logarithm of the transmittance.

■ **absorption**

1. loss or dissipation of energy as it travels through a medium, Example: radio waves lose some of their energy as they travel

through the atmosphere.
2. that portion of fiber optic attenuation resulting of conversion of optical power to heat.
absorption coefficient.
the amount of radiant energy absorbed per unit or path-length.

■ AC
abbreviation for "Alternating Current".

■ AC coupling
circuit that passes an AC signal while blocking a DC voltage.

■ AC generator
device used to transform mechanical energy into AC electrical power.

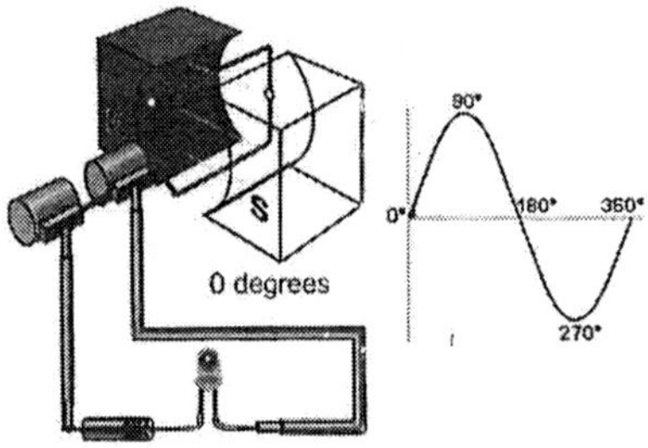

■ AC load line
a graph representing all possible combinations of AC output voltage and current for an amplifier.

■ AC power supply
power supply that delivers an AC voltage.

■ AC voltage
a voltage in which the polarity alternates.

■ AC/DC
equipment that will operate on either an AC or DC power source.

■ accelerator grid
the grid in a tube which increases electron-beam velocity.

■ accelerometer
transducer whose output is directly proportional to acceleration. Piezoelectric crystals are used to produce output.

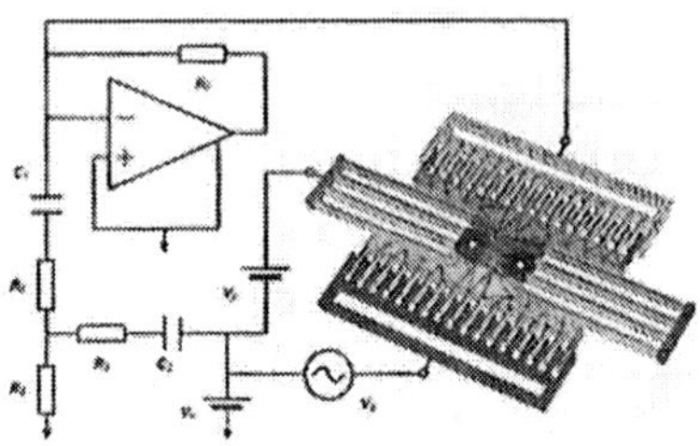

An Accelerometer Interfaces to electronic Sensing

■ acceptor atoms
1. trivalent atoms that accept free electrons from pentavalent atoms.
2. a material added as a dopant to a semiconductor to make it p-type by "accepting" valence electrons, leaving "holes" behind in the valence band that can conduct electric charge. Boron is an

acceptor dopant for silicon.

■ **accumulation region**
the region of the curve in which majority carriers are attracted to the semiconductor-insulator interface.

■ **accuracy grade**
grade of quality, determined by the maximum permissible measuring deviations within a predetermined measuring range (e.g. 1m).

■ **acquisition rate**
the rate at which the board acquires analog or digital data from an external signal input to the board. In the case of a scanning A/D, this is the aggregate conversion rate for all channels. (Number of channels times the sample rate per channel.).

■ **acquisition time (for track/hold)**
the amount of time it takes for the analog-to-digital converter (ADC) input to "acquire" the input signal. The RC time constant formed by the ADC's "track/hold" circuitry and the accuracy to which the ADC must digitise the input signal determine the length of acquisition.

■ **active component**
1. a component that changes the amplitude of a signal between input and output.
2. a component that provides gain or amplification such as transistor, integrated circuit, valve - such as a triode value.

■ **active filter**
a filter that uses an amplifier in addition to reactive components to pass or reject selected frequencies.

■ **active medium**
a medium in which lasing will take place, rather than absorption, at a given wavelength.

■ **active region**
the region of BJT operation between saturation and cut-off used for linear amplification.

■ **actuation time**
the time between application of the nominal relay coil voltage and the final closure of the relay contacts after the contact bounce interval. Also called operate time.

■ **Adaptive Differential Pulse Code Modulation (ADPCM)**
a PCM encoding technique that uses only 4 bits to encode an analog sample. By adapting the

quantising range to the difference between the two samples, it achieves the same voice quality as 8-bit PCM, but with 32 Kbps instead of 64 Kbps.

■ ADC

abbreviation for "analog to digital converter". a circuit used to convert an analog signal into a discrete digital value.

■ admittance

1. (symbol "Y") measure of how easily AC will flow through a circuit. Admittance is the reciprocal of impedance and is measured in siemens.
2. the ability of a circuit to conduct alternating current. Admittance is generally referred to as the reciprocal of impedance.
3. the reciprocal of the impedance. The admittance is the complex ratio of the current flowing through divided by the voltage across a device, circuit element, or network.

■ aerial

a length of wire or pair of conductors designed to transmit or receive radio waves.

■ AF

abbreviation for "audio frequency".

■ AF Audio Frequency

generally in the range 20Hz to 20KHz.

■ AFC

abbreviation for "automatic frequency control".

■ AFC Automatic Frequency Control

similar to Automatic Fine Tune (AFT). a circuit that keeps a receiver in tune with required transmission.

■ AGC

abbreviation for "automatic gain control".

■ air core

1. the term used to describe inductors having no iron, steel, ferrite, or other magnetic core material.
2. a narrow space between inductor cores (or between electrodes for the passage of a spark).

■ A-Law

a CEPT standard algorithm that describes the nonlinear compression performed in the analog-to-digital conversion process of PCM systems used in Europe and most other countries.

■ aliasing

1. a phenomenon which can oc-

cur whenever a signal is not sampled at greater than twice the maximum bandwidth of the signal. Causes high frequency signals to appear at low frequencies. Aliasing is minimised by filtering the signal to a bandwidth less than ½ the sample rate. When the signal starts at 0 Hz (baseband signals), bandwidth can be exchanged to maximum frequency in the definition above.

2. in sampling theory, when an input signal frequency component exceeds the Nyquist limit, the signal is "aliased" or "folded back" or replicated at other frequencies in the frequency spectrum above and below Nyquist. Normally, aliasing is due to unwanted signals beyond the Nyquist limit. To prevent aliasing, all undesired signals must be filtered adequately so that they are not digitised by the analog-to-digital converter (ADC). Aliasing can be used advantageously when under sampling.

■ alignment

1. a condition whereby the axes of machine components are coincident, parallel, or perpendicular, according to design requirements.

■ amplification factor (Synchronous)

a measure of the susceptibility of a rotor to vibration amplitude when rotational speed is equal to the rotor natural frequency (implies a flexible rotor). For imbalance type excitation, synchronous amplification factor is calculated by dividing the amplitude value at the resonant peak by the amplitude value at a speed well above resonance (as determined from a plot of synchronous response vs. rpm).

2. to tune or adjust circuitry so that performance meets specific requirements.

■ alkaline cell

a primary cell that delivers more current than a carbon-zinc cell. Also known as an "alkaline manganese cell".

■ alligator clip

spring clip on the end of a test lead used to make a temporary connection.

■ alpha

ratio of collector current to emitter current in a bipolar junction transistor (BJT). Greek letter alpha α is the symbol used.

■ alternating current

1. an electric current that rises to

a maximum in one direction, falls back to zero and then rises to a maximum in the opposite direction and then repeats.
2. current which has periodic alternations of positive and negative polarities.
3. an electric current whose direction changes direction with a frequency independent of circuit components.

■ alternation
one-half of a cycle of an alternating current or voltage.

■ alternator
a device for producing alternating current. Another name for an AC generator.

■ AM
abbreviation for "amplitude modulation"

■ American Standard Code for Information Interchange (ASCII)
the character-oriented code widely used in data processing and data communications systems.

■ ammeter
an instrument designed to measure the amplitude of current.

■ amp
ampere. The unit of electrical current. Also milliamp (one thousandth of an amp) and microamp (one millionth of an amp). One amp corresponds to the flow of about 6 x 10^{18} electrons per second.

■ ampere-turn
the product of the current flow in amperes and the number of turns of an inductor.

■ amplification
1. the process by which the amplitude of a current, voltage, or power signal is increased.
2. The growth of the radiation field in the laser resonator cavity. As the light wave bounces back and forth between the cavity mirrors, it is amp stimulated emission on each pass through the active medium.

■ amplification factor
an indication of the general amplification characteristics of a vacuum tube, defined as the ratio of a small change in plate voltage to a small change in the grid voltage, with the plate current remaining at a constant level. The symbol for amplification factor is the Greek letter ?.

■ amplifier
a device for increasing the signal voltage or signal power level.

(Classes of amplifiers are covered under 'C').

■ **amplitude**
the highest value reached by voltage, current or power during a complete cycle.

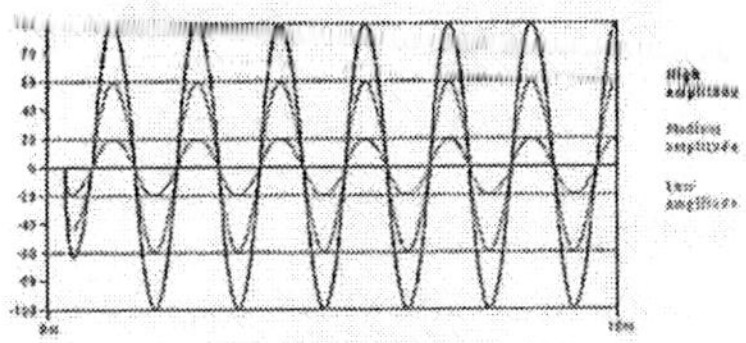

of Oscillation

■ **amplitude evaluation**
method of evaluating signals generated by dynamic scanning (with carrier frequency): the amplitude variation of two alternating voltages of the same frequency are used to determine the measuring value.

■ **amplitude modulation**
1. the type of modulation system commonly employed for radio broadcasting and picture-information broadcasting in television. The carrier signal is modulated by the low-frequency signals so that the composite waveform produced has an amplitude which varies above and below the normal carrier amplitude level at a rate and amplitude change corresponding to the modulating signal.
2. where audio signals increase and decrease the amplitude of the 'carrier wave'.

■ **amplitude shading**
a method of reducing the side lobe levels in a transducer array. The shading usually causes the main beam to broaden by applying different voltages to the elements of the array.

■ **analog**
1. information represented as continuously varying voltage or current rather than in discrete levels as opposed to digital data varying between two discrete levels.
2. a signal that varies continuously with respect to time. Conventional telephones and modems transmit analog signals. The digital or pulse output from a computer must be converted to an analog signal for transmission on these lines.

■ **analog circuit**
1. an electronic circuit in which voltages and currents vary with the intensity of an external quantity (e.g., sound level).
2. circuits that process analog signals.

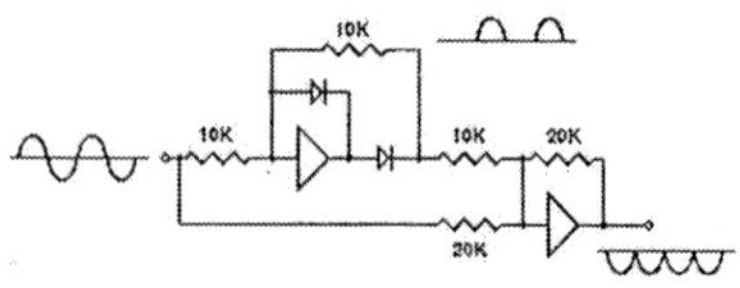

Full-wave rectifier

■ analog multiplier

a circuit that produces an output that is linearly proportional to the product of the two analog inputs.

■ analog output

an output that provides an analog signal derived from the digital information within the instrument.

■ analog ramp

a voltage output of constant slope, dV/dt (Volts/second).

■ analog trigger

an event that occurs at a user-selected point on an analog input signal. The polarity, sensitivity, and hysteresis of the analog trigger can often be programmed.

■ AND GATE

1. a logic circuit which produces an output signal when signals of the same polarity appear simultaneously at all inputs.

2. gate that produces a logic 1 when all of its inputs are 1. In all other cases the output is 0.

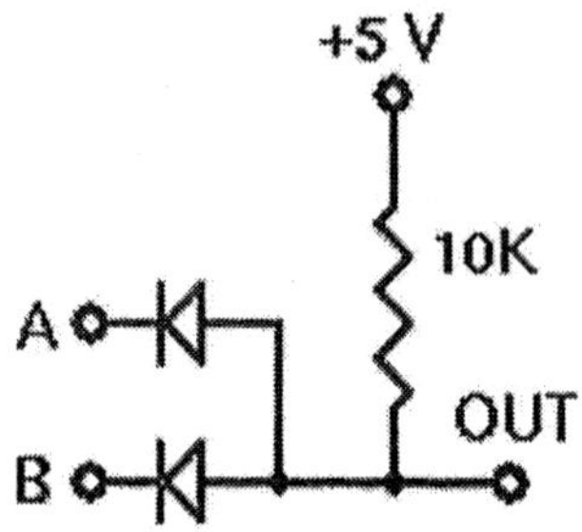

A 2 Input and Gate Schematic

■ angle encoder

angle measuring device, converts the shaft rotation angle into electrical signals (can be incremental or absolute).

■ angstrom unit

a unit of measurement for a wavelength of light (written Å), equal to one ten billionth of a meter (10^{-10} meter). Occasionally still used.

■ annealing

the process of heating a device or material in order to repair damage to the crystal lattice or activate dopant atoms.

■ anode

1. the positive electrode or terminal of a device. The "P" material of a diode.

2. The plate of a gas or vacuum tube (usually made positive) toward which electrons flow.

3. an electrical element in laser excitation which attracts electrons

from a cathode. an anode can be cooled directly by water or by radiation.

■ **ANSI**

American National Standards Institute. Accredits and implements standards Member of ISO.

■ **antenna**

1. a device for intercepting transmitted r-f signals, or for propagating r-f signals. Also called aerial.
2. a length of wire or similar that radiates (such as a transmitting antenna) or absorbs (such as a radio antenna) radio waves.
3. a device designed to radiate or absorb electromagnetic energy into space.

■ **antenna directivity**

this is the relative gain of the main beam of an antenna pattern to a reference antenna, usually an isotropic or standard dipole.

■ **antenna, receiving**

a device that converts a radiated electromagnetic wave into an electrical wave.

■ **antenna, transmitting**

a device that converts an electrical wave into an electromagnetic wave that radiates away from the antenna.

■ **anti-aliasing filter**

most commonly a low-pass filter designed to filter out frequencies higher than 1/2 the sample rate in order to minimise aliasing.

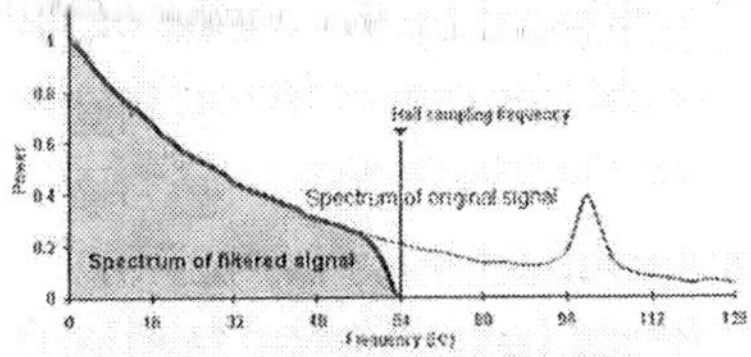

spectrum of an Anti Aliasing Filter

■ **aperture delay**

in an analog-to-digital converter (ADC), the delay in time from when the user requests the analog input to be sampled (that is, when the user puts the ADC's track/hold in "hold" mode) and the actual time when this occurs.

■ **aperture jitter**

the amount of variance in the aperture delay. This value is typically much smaller than that of the aperture delay.

■ **aperture time**

the time interval during which an amplifier or measuring instrument acquires a sample of the signal. Also known as the sample window.

■ **apparent power**

1. power attained in an AC circuit as a product of effective voltage and current which reach their peak at different times.
2. the product *EI*, in alternating-current circuits which must be multiplied by the cosine of the angle (the power factor) for obtaining true power.

■ **application specific effects (parametric testing)**

charge to breakdown of node dielectric. Change in parameter matching.

■ **application specific parameters (parametric testing)**

special parameters related only to a specific product. Term is applicable to CMOS or bipolar devices.

■ **aquadag**

the colloidal graphite coating in cathode-ray or television tubes.

■ **arc**

discharge of electricity through a gas such as lightning discharging through the atmosphere.

■ **armature**

the rotating or moving component of a magnetic circuit.

■ **armstrong oscillator**

an oscillator that uses an isolation transformer to achieve positive feedback from output to input.

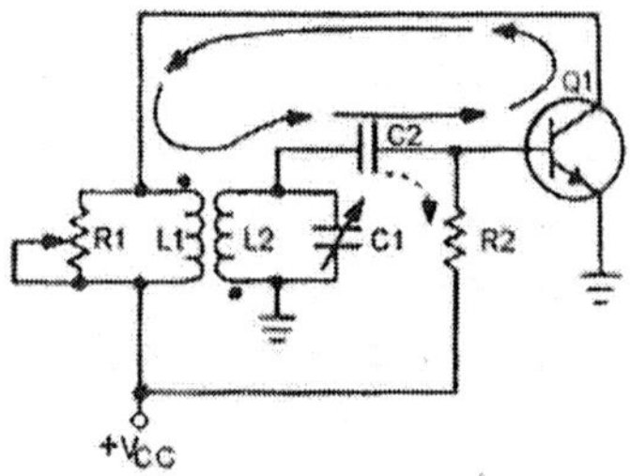

Basic Armstrong Oscillator

■ **array**

an integrated circuit fabrication technology in which the final metal interconnect layers of an otherwise pre-fabricated wafer are patterned in accordance with user-supplied design data.

■ **ASIC**

application Specific Integrated Circuit; an integrated chip which is individually custom designed for a specific application rather than a general-purpose standard chip such as a microprocessor or memory chip.

astable multivibrator

1. a circuit that has no stable state and thus oscillates at a frequency dependent on component values.
2. an oscillator that produces a square wave output from a DC voltage.

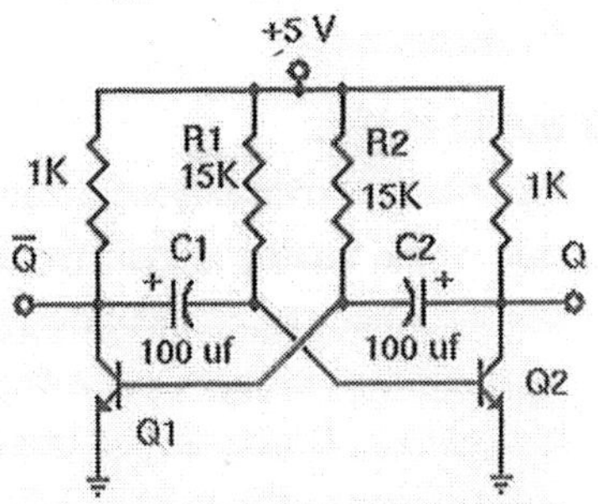

Schematic of Astable Multi Vibrator

asymmetrical support

rotor support system that does not provide uniform restraint in all radial directions. This is typical for most heavy industrial machinery where stiffness in one plane may be substantially different than stiffness in the perpendicular plane. Occurs in bearings by design, or from preloads such as gravity or misalignment.

asynchronous

1. vibration components that are not related to rotating speed .
2. in hardware, it is an event that occurs independent of other events; it is not synchronised with a clock signal. In software, it refers to a function that begins an operation and returns to the calling program prior to the completion or termination of the operation.

asynchronouc mode

a standard for data transmission where each data package has a start and stop bit.

asynchronous transmission

in data transmission, each character is individually synchronised, usually by start bits and stop bits. Timing information is usually included in the transmitted character.

ATE

Automatic Test Equipment; computer controlled equipment used in the production testing of packaged ICs. Test voltage sequences (test pattern) are applied and responses compared to data on file or to a known-to-be-good IC.

ATM

Asynchronous Transfer Mode. a high-bandwidth, low- connection-oriented switching and multiplexing technique using fixed-sized cells.

attenuate

1. to reduce the amplitude of an action or signal. The opposite of amplification.
2. the dissipation of the power of a transmitted signal as it travels over a wire.
3. a decrease in signal magnitude between two points. These points may be along a radio path, transmission line or other device.
4. The reduction in optical power as it passes along a fiber, usually expressed in decibels (dB).

attenuator

1. a circuit for reducing signal amplitude by the use of either fixed or variable components.
2. a device that reduces signal power in a fiber optic link by inducing loss.

attitude angle (steady-state)

the angle between the direction of steady-state preload through the bearing centreline, and a line drawn between the shaft centreline and the bearing centreline. (Applies to fluid-film bearings.)

auto spectrum (Power Spectrum)-DS

a spectrum display whose magnitude represents the power at each frequency, and which has no phase.

audio

relating to frequencies that can be heard by the human ear. Approximately 20 Hz. to 20 kHz.

audio amplification

the increase of signal amplitudes within the 16 cps to 20,000 cps frequency range.

audio codec

the critical interface between outside world analog signals (such as the human voice) and the digital data processing inside a mobile phone. It acts as the main contributor to the voice quality of a mobile phone. It converts the digital signal received from the baseband subsystem into an analog signal that is fed to the loudspeaker and also converts the analog signal from the microphone into a digital signal.

audio codec ASICs

ASICs designed to perform the Audio CODEC function.

audio crosspoint switch

a crosspoint that isn't high performance, but is capable of handling the larger signals common to audio applications.

auto calibration
the process a PAC chip goes through to correct certain inaccuracies inherent in the part (such as offset). The result is a part that's much more accurate because local circumstances (temperature, age) can be compensated for.

autodyne
a system of heterodyne reception. The circuit combines the functions of an oscillator and detector.

automatic gain control
1. abbreviated as AGC . a circuit that adjusts the gain of a stage so that the volume is constant even though the input signal may vary over a wide range.
2. the circuit samples demodulated gain levels and provides an automatic correction bias for maintaining a predetermined signal amplitude.

auto polarity
the ability of an instrument to measure and display an input of either polarity without switching the input leads.

auto ranging time
for instruments with auto ranging capability, the time interval between application of a step input signal and its display, including the time for determining and changing to the correct range.

auto ranging
the ability of an instrument to switch among ranges automatically. The ranges are usually in decade steps.

autotransformer
a single winding transformer where the output is taken from taps on the winding.

avalanche breakdown
a breakdown that is caused by the cumulative multiplication of charge carriers through field-induced impact ionisation.

Avc
abbreviation for "automatic volume control"

average power
the average over time of a modulated signal.

average responding
1. a measurement where the displayed value is proportional to the average of the absolute values of all input waveforms within a specified frequency range. It is calibrated in the rms value of a sine wave.
2. a value of voltage or current where the area of the wave

above the value equals the area of the wave below the value.

■ average value

the average of all instantaneous values of I or E taken for a half cycle of a sine wave. It is the product of 0.636 times the peak amplitude.

■ averaging

in a DSA, digitally averaging several measurements to improve accuracy or to reduce the level of asynchronous components. Refer to definitions of rms, time, and peak-hold averaging.

■ avionics

aviation electronics.

■ AWG

abbreviation for "American Wire Wauge". a gauge that assigns a number value to the diameter of a wire.

■ axial

in the same direction as the shaft centreline.

■ Axial Position

the average position, or change in position, of a rotor in the axial direction with respect to some fixed reference position. Ideally the reference is a known position within the thrust bearing axial clearance or float zone, and the measurement is made with a displacement transducer observing the thrust collar.

■ axial flow laser

the simplest and most efficient of the gas lasers. an axial flow of gas is maintained through the tube to replace those gas molecules depleted by the electrical discharge used to excite the gas molecules to the lasing state.

■ back emf

back electromotive force. (Also called counter emf on occasion.) The voltage generated in an inductor when alternating current flows through it. The counter emf has a polarity opposite to the source voltage and it opposes a current change.

■ back end

back-end refers to the package assembly and test stages of production in semiconductor manufacturing. It includes burn-in and environmental test functions.

■ back reflection, optical return loss

light reflected from the cleaved or polished end of a fiber caused by the difference of refractive indices of air and glass. Typically 4% of the incident light. Expressed in dB relative to incident power.

■ **back-end assembly**
the second phase of chip manufacturing during which the die is assembled into packaging designed not only to protect it, but also to provide external connections via a series of very fine wires.

■ **back-end design**
design activities starting with either a net list or hardware description language version of a chip design and ending with the layout (physical design) of the chip.

■ **background task**
an operation that can take place while another program or processing routine is running without apparent interruption to that program or routine. For example, an interrupt or DMA operation.

■ **backscattering**
the scattering of light in a fiber back toward the source.

■ **balanced bridge**
condition that occurs when a bridge circuit is adjusted to produce a zero output.

■ **band bending**
the increase or decrease of the energy band at the interface between materials (such as semiconductor/insulator). Carriers will experience an electric field where the energy band bends and fall down or float up, depending on type.

■ **band stop filter**
1. an electronic network which passes signals having frequencies above and below the band-elimination frequency span.
2. a tuned circuit designed to stop frequencies between a lower cut-off frequency (f1) and a higher cut-off frequency (f2) of the amplifier while passing all other frequencies.

■ **band pass Filter**
1. an electronic network which transfers through it a specific band of signal frequencies.
2. a tuned circuit designed to pass a band of frequencies between a lower cut-off frequency (f1) and a higher cut-off frequency (f2). Frequencies above and below the pass band are heavily attenuated.

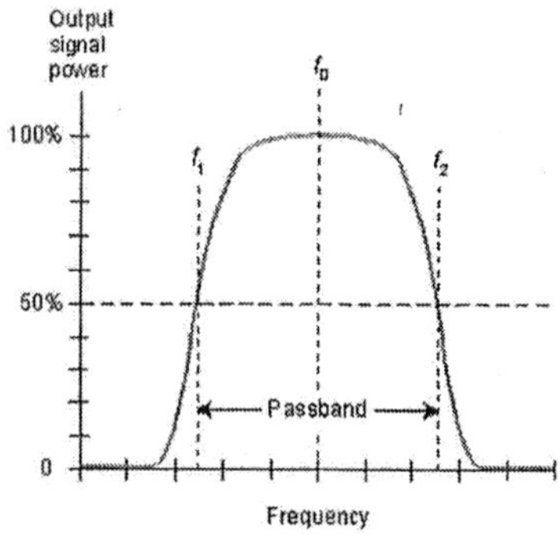

Response of a Band Pass Filter

3. a filter that allows only signals between a lower and a higher frequency limit to pass through. All other ("out-of-band") frequencies are attenuated.

4. a filter with a single transmission band extending from lower to upper cutoff frequencies. The width of the band is normally determined by the separation of frequencies at which amplitude is attenuated by 3 dB (a factor 0.707).

■ bandwidth

1. width of the band of frequencies between the half power points.

2. the range of signal frequencies or bit rate within which a fiber optic component, link or network will operate.

3. the carrying capacity or sise of a communications channel; usually expressed in hertz (cycles per second) for analog circuits (the original meaning of the term) or in bits per second (bps) for digital circuits (newer meaning). The difference between the highest and lowest frequencies of a band that can be passed by a transmission medium without undue distortion.

4. usable frequency range of the antenna, measured in MHz ($1x10^5$ cycles per second, Mega-Hertz) or GHz ($1x10^9$ cycles per second, Giga-Hertz).

■ bank

a group of relays with a common connection for scanning or multiplexing applications.

■ barrier potential

the natural difference of potential that exists across a forward biased pn junction.

■ base

1. one of the pins of a TRANSISTOR. This pin is the 'trigger' which will switch the transistor on when an electric current, usually greater then 0.6 voltage, is applied. If there is little or no current, the transistor is off. The base is also known as the GATE.

2. the region that lies between the emitter and collector of a

bipolar junction transistor (BJT).

base address
an I/O address that is the starting address for programmable registers. All subsequent registers are accessed by adding to the base address.

base biasing
a method of biasing a BJT in which the bias voltage is supplied to the base by means of a resistor.

base terminal
that electrode of a transistor which compares generally to the grid of a vacuum tube.

baseband
the frequency band occupied by the aggregate of all the voice and data signals used to modulate a radio carrier.

baseband processing subsystem
the manner in which a microcontroller and a DSP control a baseband processor and interact with the operator of a mobile phone through the phone's display and keypad.

baud
a unit of signaling speed equivalent to the number of signaling elements per second. Modems that transmit faster than 1800 bps send multiple bits per baud. For example, 9600 bps modems transmit 4 bits per baud.

beam bender
hardware assembly or optical device, such as a mirror, capable of changing laser beam direction; used to re-point the beam and in "folded," compact delivery systems.

beam diameter
the diameter of that portion of the beam which contains 86% of the output power.

beam expander
optical device increasing beam diameter and reducing divergence. Result a smaller focused spot for more distance between lens and part.

beam pattern
beam patterns show the relative amplitude of the acoustic pressure (generated or received) as a function of direction relative to the transducer. For reciprocal transducers the transmit and receive beam patterns are basically the same. Beam patterns are three-dimensional.

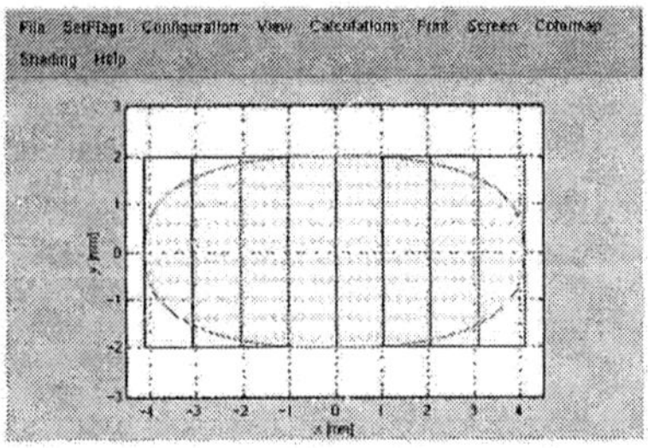

■ **beam splitting**
optically splitting a laser beam into two or more beams, allowing work on more than one side of a part at the same time—but at somewhat less power than with a multiple-output beam system.

■ **beam steering**
the method of steering the main lobe of a transducer to a certain direction.

■ **beam width**
1. the width of the main beam lobe, in degrees, of the transducer. It is usually defined as the width between the "half power point" or "-3dB" point.
2. The amount of angle (either azimuth or zenith) covered by the antennas maximum gain (in decibels) until the signal drops of 3dB in each direction.

■ **bearer channel**
a 64 Kbps channel in Basic Rate and Primary Rate ISDN services used to transmit the customer's voice, data, and video information.

■ **beat frequency**
the signal produced as a result of heterodyning two signals of different frequency.

■ **bel**
the unit of the difference between two powers. The general expression is the decibel (one-tenth of a bel).

■ **Bending loss, microbending loss**
loss in fiber caused by stress on the fiber bent around a restrictive radius.

■ **beta**
the ratio of the collector current to the base current of a bipolar transistor, commonly referred to as either the common-emitter current gain or the current amplification factor.

■ **BFO**
Beat-frequency oscillator.

■ **bias**
1. a DC voltage applied to a device to control its operation.
2. the potential applied between the grid and cathode of a tube or between transistor elements to

provide an operating point at zero signal input.

bias temperature stress (BTS) cycle

the stress cycle consists of a bias voltage applied while the device is heated to a stress temperature for a duration called the stress time.

bias temperature stress (BTS) test

a test for mobile ion drift consisting of C-V curves alternated with positive and negative voltage BTS cycles.

bias voltage

a voltage applied to a circuit or device to establish a reference level or operating point of the device during testing.

BICMOS

Bipolar Complementary Metal Oxide Semiconductor. an IC technology combining the linearity and speed advantages of bipolar and the low-power advantages of CMOS on a single IC. BiCMOS can operate at either ECL (emitter-coupled-logic) or TTL (transistor-transistor-logic) levels, and is ideal for mixed-signal devices.

binary

a number system having only two symbols, 0 and 1. a base 2 number system.

binary coding (unipolar)

a digital coding scheme in which zero is represented by all zeros (00000000) and full scale by all ones (11111111).

binary synchronous communications (Bisync)

an IBM communications protocol that has become an industry standard. It uses a defined set of control characters and control character sequences for synchronised transmission of binary coded data between stations in a data communications system. Bisync is a byte-oriented protocol.

bipolar

a signal that swings both above and below analog ground, thus having positive and negative values.

bipolar junction transistor (BJT)

a three terminal device in which emitter to collector current is controlled by base current.

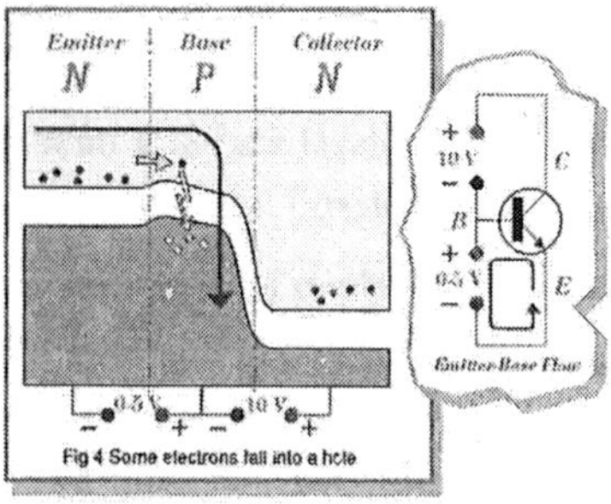

Fig 4 Some electrons fall into a hole

■ **bipolar return to zero (BRZ)**
a bipolar signal in which each pulse returns to zero amplitude before its time period ends. This prevents the buildup of DC current on the signal line.

■ **bipolar violation (BPV)**
the occurrence of two successive pulses of the same polarity in a bipolar signal.

■ **bistable**
circuit that has two stable states.

■ **bistable multivibrator**
a multivibrator with two stable states. an external signal is required to change the output from one state to the other. Also called a latch.

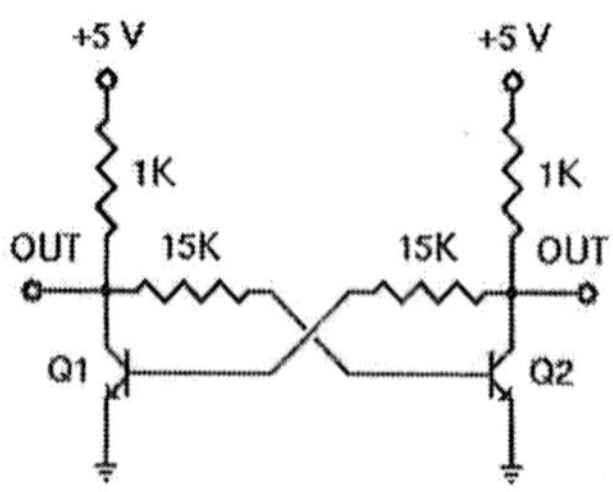

Schematic Diagram of Bistable Multivibrator

■ **bit**
originally a contraction of "Binary digit." A bit is the basic unit in data communications and the smallest element of information in the digital system. an individual bit is either a 1 or a 0. Computer hardware stores and manipulates information by grouping many bits together.

■ **bit error ratio**
the ratio of bits with errors to the total number of bits detected, usually expressed as a number with an exponent to a power of ten. Used to measure the quality of a signal path. Measure of transmission quality indicating the number of bits incorrectly transmitted in a given bit stream compared to the total number of bits transmitted in a given duration of time.

■ **bit robbing**
a technique used in Tl multiplexing whereby the least significant bit (bit eight) of each byte in selected frames is "robbed" from use in carrying message information, and instead is used to carry signaling information.

■ **bit-error rate (BER)**
the fraction of data bits transmitted that are received in error.

■ **bit-maps (parametric testing)**
a yield parameter. Number and location of defective bits in DRAM or SRAM sub-arrays.

■ **blade passing frequency**
a potential vibration frequency on any bladed machine (turbine, axlal compressor, fan, etc.). It is represented by the number of blades times shaft- rotating frequency.

■ **blanking**
electron-beam cutoff in a cathode-ray tube during beam retrace.

■ **blanking distance**
minimum sensing range in an ultrasonic proximity sensor. Blanking distance is a function of the ring down time of the transducer as the transducer must ring down before it can receive the sound reflected from the target.

■ **bleeder**
one or more resistors shunting the output of a power supply for improving voltage regulation by providing a fixed current drain.

■ **bleeder current**
a current drawn continuously from a source. Bleeder current is used to stabilise the output voltage of a source.

■ **block size**
the number of samples used in a DSA to compute the Fast Fourier Transform. Also the number of samples in a DSA time display. Most DSAs use a block sise of 1024. Smaller block sise reduces frequency resolution.

■ **blocking oscillator**
a relaxation oscillator with periodic grid (or base) cutoff (blocking) as its basic characteristic.

■ **bluetooth**
a radio technology designed to standardise the transmission of signals over short distances between telephone, computers and other devices without the use of wires.

■ **bode**
rectangular coordinate plot of lx component amplitude and phase (relative to a key phasor) vs. running speed.

■ **bode plot**
a graph of gain versus frequency.

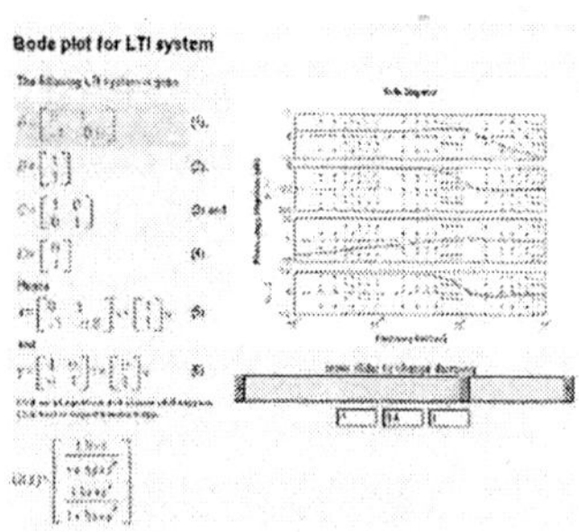

bottoming
a transistor in the fully conducting state.

bow
a shaft condition such that the geometric centreline of the shaft is not straight.

branch current
the portion of total current flowing in one path of a parallel circuit.

break-before-make
disconnecting the present circuit before connecting a new circuit.

breakdown voltage
voltage at which the breakdown of a dielectric or insulator occurs.

breakover voltage
minimum voltage required to cause a DIAC to break down and conduct.

brewster windows
the transmissive end (or both ends) of the laser tube, made of transparent optical material and set at Brewster's angle in gas lasers to achieve zero reflective loss of vertically polarised light. Non-standard on industrial lasers, but a must if polarisation is desired.

bridge circuit
a circuit usually diagrammed as a square formation of components which gives a null indication on a sensing device when a balance is achieved.

bridge rectifier
a circuit using four diodes to provide full wave rectification. Converts an AC voltage to a pulsating DC voltage.

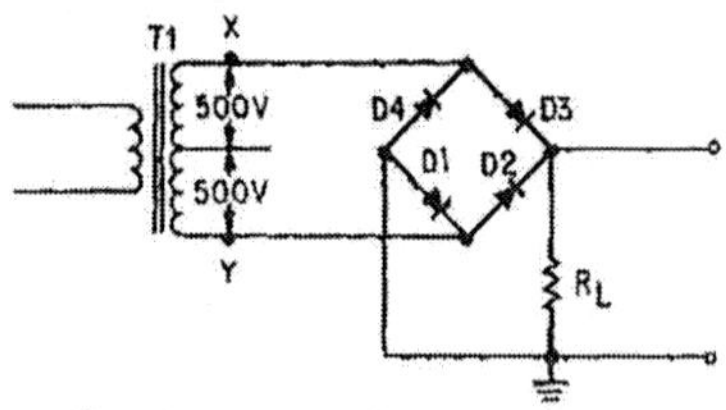

B FULL-WAVE BRIDGE RECTIFIER

Circuit Diagram of Bridge Rectifier

bridged tap
any part of the local loop that is not in the direct transmission path between the input and the service user.

broadband
refers to a communications network in which frequency range is divided into multiple indepen-

dent channels for simultaneous transmission of signals such as voice, data or video.

broadcast band
the term generally applied to the r-f frequency span between 550 kHz and 1600 kHz allocated to radio transmission.

buffer
1. an amplifier used to isolate a load from a source.
2. a protective coating applied directly on the fiber.
3. a circuit used to drive a heavy load. It is often implied that the gain of a buffer is one.

buffer memory
temporary storage area for acquired or generated data.

buffer stage
an amplifier or other circuit usually employed in r-f amplifier (transmitter) stages for effective isolation between the oscillator circuit and subsequent Class C amplifiers.

bulk resistance
the natural resistance of a "P" type or "N" type semiconductor material.

burn-in
the operation of items prior to their ultimate application, intended to stabilise their characteristics and identify early failures.

burst mode
a data acquisition mode in which a group of analog input channels are scanned at an interval determined by the pacer clock and the signal from each channel within the scan is converted at a higher rate determined by the burst mode conversion clock. This mode minimises the skew between channels.

bus
an interconnection system that allows each part of a computer to communicate with the other parts.

Butterworth Filter
a type of active filter characterized by a constant gain (flat response) across the mid band of the circuit and a 20 dB per decade roll-off rate for each pole contained in the circuit.

BW
abbreviation for bandwidth.

bypass capacitor
1. a capacitor so wired in a circuit that it has low reactance for signal frequencies and, hence, effectively shunts them around another component.

2. a capacitor that provides a path of low impedance - low resistance to AC signals.

3. a capacitor used to provide an AC ground at some point in a circuit.

byte

a group of binary digits that combine to make a word. Generally 8 bits. Half byte is called a nibble. Large computers use 16 bits and 32 bits.

cable

one or more fibers enclosed in protective coverings and strength members.

cable correction using reference capacitors

in this method, precise capacitance references are measured through the connecting cables to determine their S-parameters. The instrument uses the S-parameters to correct for cable errors.

cable driver

an amplifier, usually in a gain of 2, suitable for driving the low resistance of a double terminated cable (load resistance = 150 ohms for video, 100 ohms for instrumentation).

cable plant

the combination of fiber optic cable sections, connectors and splices forming the optical path between two terminal devices.

CAD

abbreviation for "computer aided design" the use of computer programs to help create integrated circuits and systems.

calibration

1. a test during which known values of the measured variable are applied to the transducer or readout instrument, and output readings varied or adjusted.

2. to adjust the correct value of a reading by comparison to a standard.

calorimeter

an instrument which measures the heat generated by absorption of the laser beam—another way to measure laser power.

Campbell Diagram

a mathematically constructed diagram used to check for coincidence of vibration sources(i.e. 1 x imbalance, 2 x misalignment) with rotor natural resonances. The form of the diagram is like a spectral map (frequency versus rpm),but the amplitude is represented by a rectangular plot, the larger the amplitude the larger the rectangle. Also known as an interference diagram.

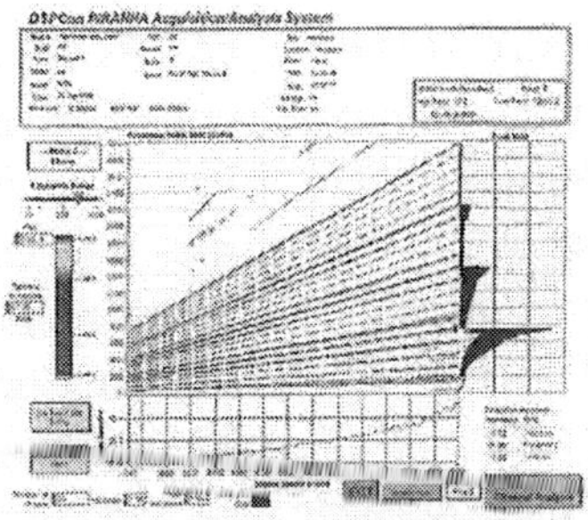

An Example of Cambell Diagram

CAP

Carrierless Amplitude and Phase Modulation. a transmission technology for implementing a DSL connection. Transmit and receive signals are modulated into two wide- bands using passband modulation techniques.

capacitance

1) the ability of a capacitor to store an electrical charge. The basic unit of capacitance is the Farad.

2) letter C, units Farad F, microfarad ìF, nanofarad nF, picofarad µF. a capacitor has capacitance. The property that allows a component to store electric charge when a potential difference exists between its terminals. Stray capacitance - unwanted capacitance - created by the proximity of other components.

capacitance meter

any meter that measures capacitance. Most capacitance meters do not have test signal levels appropriate for C-V testing and do not contain a staircase voltage source.

capacitance non-linearity

a relationship between the actual and measured capacitance across a range that cannot be described with a straight line. The most common source of non-linearity is connection or cable error..

capacitance voltage

the relationship (or curve) of capacitance versus voltage for a device. Generally measured on Schottky, P-N, or MIS structures. C-V curves are of interest to ascertain device geometry, internal device potentials, and device charges.

capacitance

in a capacitor or system of conductors and dielectrics, the property that permits the storage of electrically separated charges when potential differences exist between the conductors. Capacitance is related to charge and voltage as follows

$C = Q/V$, where C is the capacitance in farads, Q is the charge in coulombs, and V is the voltage in volts.

capacitive displacement current
the current that flows in a capacitor in response to a voltage change across the capacitor.

capacitive load drive
the maximum capacitance a digital-to-analog converter can drive without generating extraneous outputs.

capacitive reactance
the opposition to current flow provided by a capacitor. Capacitive reactance is measured in ohms and varies inversely with frequency.

capacitor microphone
microphone whose operation depends on variations in capacitance caused by varying air pressure on the movable plate of a capacitor.

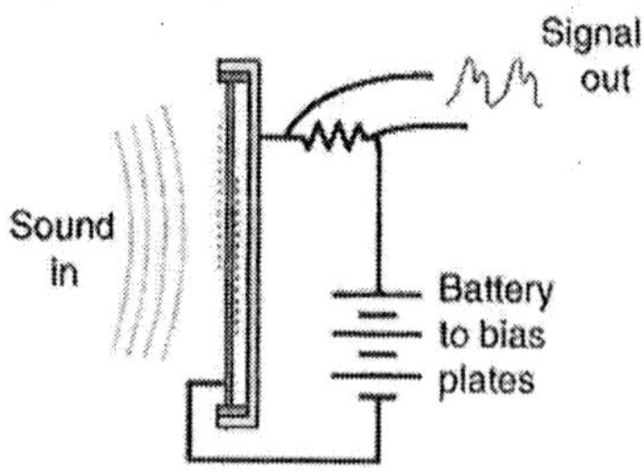

Condenser

capacitor
a capacitor can store small amounts of electrical charge. It consists of two ELECTRODES with a non-conducting layer in between them. They come in many different shapes. The capacitors in the picture are radial. This means that the legs come out of the bottom of the can. Axial capacitors have their legs coming out of either end. They are passive devices.

carbon microphone
microphone whose operation depends on pressure variation in carbon granules causing a change in resistance.

carbon resistor
resistor of fixed value made by mixing carbon granules with a binder which is moulded and then baked.

carbon-film resistor
device made by depositing a thin carbon film on a ceramic form.

carrier
a signal capable of being modu-

lated to carry information.

■ **carrier concentration**
the concentration in cm^{-3} of free carriers. This is essentially a synonym for doping concentration. Strictly speaking, though, electrical techniques such as C-V measure carrier concentration because it is the electrical carriers we detect, not the chemical presence of dopant ions.

■ **carrier frequency method**
scanning method used mainly with magnetic and inductive measuring
systems (see dynamic scanning).

■ **carry current**
the maximum continuous current of closed relay contacts. Most relays are rated higher for carry current than switched current.

■ **cascade**
1. circuits connected in sequence where the anode of one stage is coupled to the grid of the next stage.
2. a method of connecting circuits in series so that the output of one is the input of the next.

■ **cascaded amplifier**
an amplifier with two or more stages arranged in a series configuration.

■ **cascode**
circuits connected in sequence where the anode of one stage is coupled to the cathode of the next stage.

■ **cascode amplifier**
a high frequency amplifier made up of a common-source amplifier with a common-gate amplifier in its drain network.

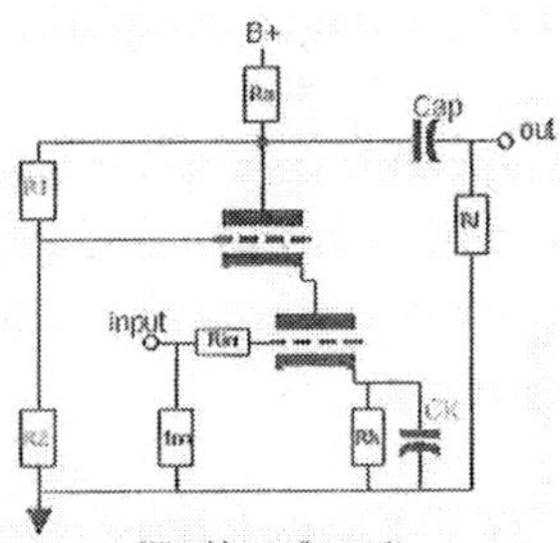

Ultra-Linear Cascode

■ **cathode**
1. the negative terminal electrode of a device. The "N" material in a junction diode.
2. the element within a gas or vacuum tube which emits electrons.
3. the element providing the electrons for the electrical discharge used to excite the lasing medium.

■ **cathode follower**
a tube circuit which develops the output signal across the cathode circuit, with the anode at signal ground. an impedance step-down is procured. (Comparable

to the grounded-collector transistor circuit.).

■ **cathode ray oscilloscope**

an instrument with a screen that provides a visual indication of the voltage wave form at any point in a circuit.

■ **cathode ray tube (CRT)**

1. vacuum tube used to display data in a visual form. Picture tube of a television or computer terminal.

2. a tube with a phosphor-treated face-plate wherein an electron beam traces an image by causing fluorescence.

CRT Components

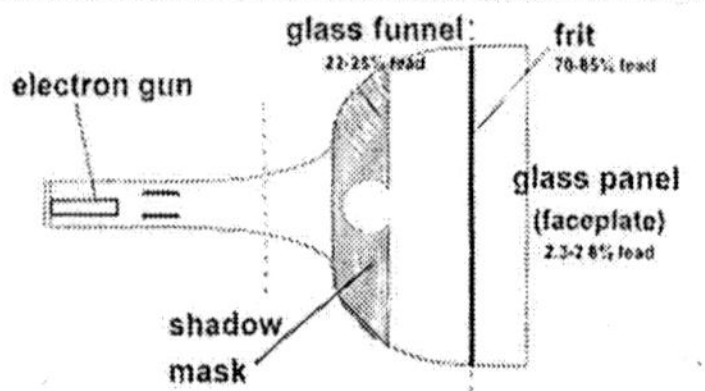

Components of CRT

■ **cavitation**

a condition which can occur in liquid-handling machinery (e.g. centrifugal pumps) where a system pressure decrease in the suction line and pump inlet lowers fluid pressure and vaporisation occurs. The result is mixed flow which may produce vibration.

■ **cell library**

a set of computer data files which contain all the information needed to model, simulate, and layout pre-designed integrated circuit modules which can then be automatically interconnected.

■ **centre frequency**

1. frequency to which an amplifier is tuned. The frequency half way between the cut-off frequencies of a tuned circuit.

2. the term usually applied to the unmodulated FM carrier frequency.

3. for a bandpass filter, the centre of the transmission band, measured in a linear scale.

■ **centre tap**

midway connection between the two ends of a winding.

■ **centre tapped rectifier**

circuit that make use of a centre tapped transformer and two diodes to provide full wave rectification.

■ **centre tapped transformer**

a transformer with a connection at the electrical centre of a winding.

■ **ceramic capacitor**

generally a single layer capacitor

that is flat and has a brown coating, Also have the name monoblock or monolithic in which the capacitor is made even smaller by creating multi-layers and coated in orange or blue paint

■ ceramic transducers
a ceramic device that converts energy from one form to another by taking physical phenomena such as pressure or temperature, and converting them into an electronic signal and vice versa, e.g. as a loudspeaker.

■ channel (semiconductors)
the conducting region between a FET source and drain that is controlled by the applied gate voltage. The majority carriers are holes in P-channel FETs and electrons in N-channel FETs.

■ channel (switching)
one of several signal paths on a switching card. For scanner or multiplex cards, the channel is used as a switched input in measuring circuits or as a switched output in sourcing circuits. For switch cards, each channel's signal paths are independent of other channels'. For matrix cards, a channel is established by the actuation of a relay at a row and column cross point.

■ channel isolation
on a switching card, the isolation from signal high and low of one channel to signal high and low of any other channel (or the output on switch or scanner cards). Specified as resistance and capacitance, except for RF cards (decibels and frequency range).

■ channel-gain queue
a user-defined sequence of channels with corresponding gains.

■ characteristic curves
graphs of either tube or transistor characteristics of voltage, current, and bias factors.

■ characteristic impedance
the iterative (surge) impedance of a transmission line or an attenuator network equivalent to that which would prevail for a line of infinite length.

■ characterisation
the process of collecting measured electrical data about the operation of an integrated circuit component or module to enable it to be accurately modeled in a computer simulation.

■ charge amplifier
amplifier used to convert accelerometer output impedance

from high to low, making calibration much less dependent on cable capacitance.

■ charge carrier

a carrier of electric charge. an electron carries negative charge and a hole carries positive charge. Charge carriers are free to move through the semiconductor device.

■ charge coupled device (CCD)

a semiconductor device, often used for sensing light, which operates by storing charge on capacitors and selectively moving that charge through the device by manipulating voltages on its electrodes.

■ Chebyshev filter

a type of active filter characterized by high roll-off rates (40 dB per decade per pole) and midband gain that is not constant.

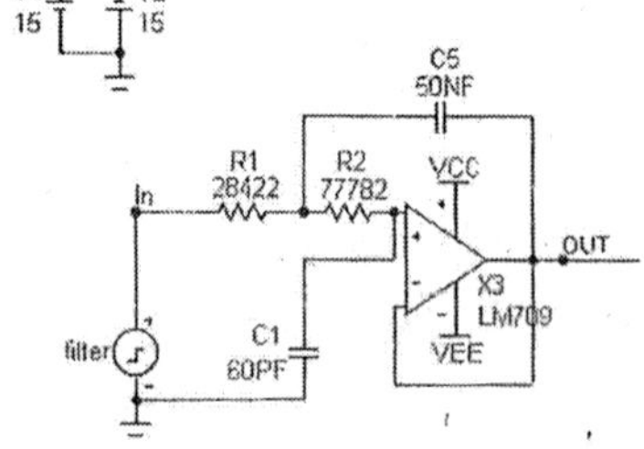

■ chip

1. another name for Integrated Circuit or the piece of silicon on which semiconductors are created. 2. a single die cut from a wafer which usually contains a single integrated circuit.

■ choke coil

an inductor so designed that it has a high reactance when placed in series with a signal-carrying lead.

■ chopper

a circuit which converts dc to ac by periodic interruption (chopping) of the DC.

■ chromatic dispersion

1. the temporal spreading of a pulse in an optical wave guide caused by the wavelength dependence of the velocities of light. 2. the temporal spreading of a pulse in an optical wave guide caused by the wavelength dependence of the velocities of light.

■ circuit

interconnection of components to provide an electrical path between two or more components.

■ circuit breaker

a protective device used to open a circuit when current exceeds a maximum value. In effect a reusable fuse.

■ **cladding**

the lower refractive index optical coating over the core of the fiber that "traps" light into the core.

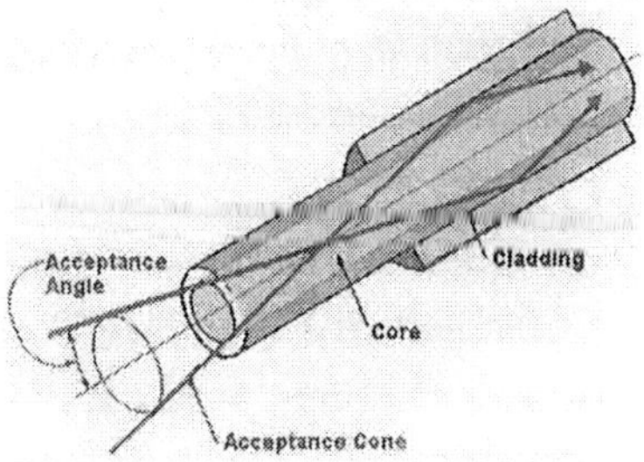

Cross Sectional View of Fiber

■ **connector**

a device that provides for a demountable connection between two fibers or a fiber and an active device and provides protection for the fiber.

■ **clamper**

1. a diode circuit used to change the DC level of a waveform without distorting the waveform.
2. a circuit designed to restore the d-c component of a signal waveform.

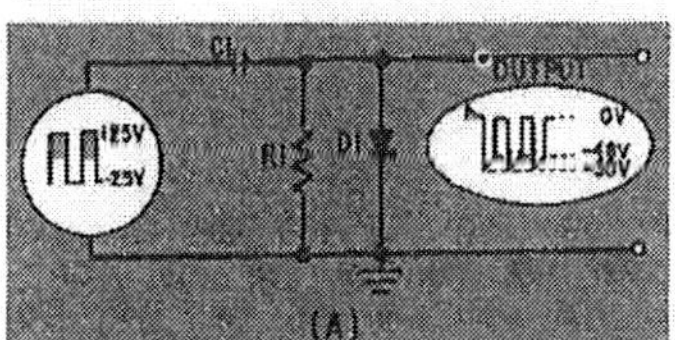

Negative Clamper Circuit and Waveform

■ **clapp oscillator**

a variation of the Colpitts oscillator. an added capacitor is used to eliminate the effects of stray capacitance on the operation of the basic Colpitts oscillator.

■ **class A amplifier**

1. a linear amplifier biased so the active device conducts through 360 degrees of the input waveform.
2. an amplifier so designed that the bias is set on the linear portion of the characteristic curve and in which the input signal amplitude is confined to the linear operating region.

■ **class AB1 Amplifier**

an amplifier with slightly higher bias than the Class a and with a larger input signal having amplitudes reaching (but not exceeding) the cutoff and positive regions of grid-signal swing.

■ **class AB2 Amplifier**

an amplifier with higher bias than the Class AB1 and with grid signal excursions extending into the cutoff and positive tube characteristic regions.

■ **class B amplifier**

1. an amplifier with two active devices. The active components are biased so that each conducts for approximately 180 degrees of the input waveform cycle.
2. an amplifier with bias at or near

the tube cutoff point. Positive alternations of the input signal cause current flow.

■ **class C amplifier**

1. an amplifier in which the active device conducts for less than 180 degrees of the input waveform cycle.

2. an amplifier biased beyond the cutoff point so current flows for only a portion of the positive alternations of the input signal.

■ **clipper**

diode circuit used to eliminate part of a waveform.

■ **clipper**

a circuit designed to remove portions of the input signal amplitudes.

■ **clipping**

distortion caused by overdriving an amplifier.

■ **clock**

a square waveform used for synchronizing and timing of several circuits.

■ **closed circuit**

circuit having a complete path for current flow.

■ **closed loop buffer amplifier**

an op amp configured for a gain of 1 or 2, with the feedback network on the IC, thus saving board space .

■ **closed-loop gain**

gain of an amplifier when a feedback path is present.

■ **CMOS**

Complimentary Metal-Oxide-Semiconductor, Family of logic devices that uses p-type and n-type channel devices on the same IC. It has the advantage of offering medium speed and very low power requirements.

CMOS

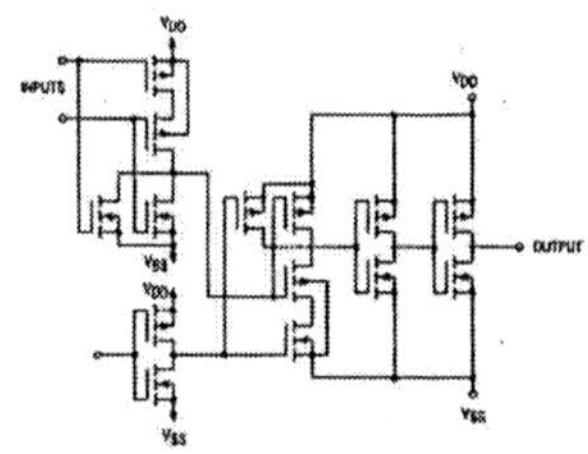

Complementary Metal-Oxide-Semiconductor

■ **CO_2 LASER**

a laser largely used in industry in which the primary lasing medium is carbon dioxide.

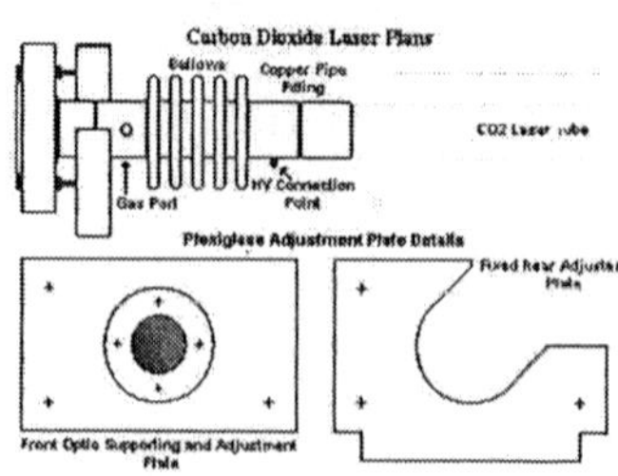

coax

1. coaxial cable. a round cable with a central conductor and screening around with an insulating medium between.

2. coaxial cable, a centre conducting wire separated from an outer braid conductor by a dielectric (insulator). The braid surrounds the dielectric and when connected to ground, prevents (shields) the centre conductor from radiating energy. The spacing between conductors, composition of the dielectric and sises of the conductors determine the impedance of the cable. Coax is used to transfer radio frequency energy from the transmitter to the antenna.

coaxial cable

transmission line in which the signal carrying conductor is covered by a dielectric and another conductor.

coaxial gas

most laser welding is done with a shield of inert gas flowing over the work surface to prevent plasma oxidation and absorption, to blow away debris, and to control heat reaction. The gas jet has the same axis as the beam so the two can be aimed together.

CODEC

a coding/decoding device that converts, or encodes, analog signals into a form for transmission on a digital circuit. The digital signal is then decoded back to analog signals at the receiving end of the transmission link. CODECs allow voice and video transmission over digital links and may also support signal compression.

coefficient of coupling

the degree of coupling between two circuits.

coercive force

magnetizing force needed to reduce residual magnetism in a material to zero.

coherence

measures how much of the output signal is dependent on the input signal in a linear and time-invariant way. It is an effective means of determining the similarity of vibration at two locations, giving insight into the possibility of cause and effect relationships.

coherent light

radiation composed of wave trains vibrating in phase with each other. Simply expressed parallel rays of light.

■ coil

a conductor wound in a series of turns.

■ coil resistance

a nominal value of the resistance of a relay coil winding at a specified ambient temperature.

■ cold junction

the junction in a thermocouple circuit that is held at a stable known temperature. Also known as reference junction.

■ cold switching

closing the relay contacts before applying voltage and current and removing voltage and current before opening the contacts. (Contacts do not make or break current).

■ cold-cathode tube

a tube which requires no external current source for producing electron emission by heat. Other means initiate electron flow.

■ cold-junction compensation

a method of compensating for ambient temperature variations in thermocouple circuits.

■ collector

the semiconductor region in a bipolar junction transistor through which a flow of charge carriers leaves the base region.

■ collector characteristic curve

a graph of collector voltage over collector current for a given base current.

■ collimated light

divergent light rays rendered parallel by means of a lens or other device, allowing a sharp image of the object to be focused at the focal plane of the lens.

■ colour code

1. set of colours used to indicate value of a component.

2. the identification of electronic components by colour bands or dots related to numerical or alphabetical values.

■ Colpitts oscillator

1. an oscillator with a pair of tapped capacitors in the feedback network.

2. an oscillator characterized by series capacitors shunting the resonant circuit inductor.

COLPITTS OSCILLATOR SYSTEMS

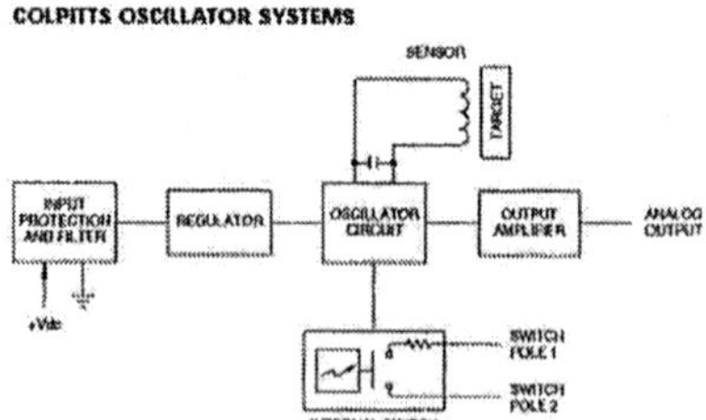

Colpitts Oscillator Block Diagram

■ **column**

as viewed on the schematic of a matrix relay card, the vertical signal lines or a vertical group of relays.

■ **combustible liquid**

materials with a flash point above 37.8°C (100°F).

■ **common base amplifier**

a BJT circuit in which the base connection is common to both input and output.

■ **common base connection**

same as ground base connection. a mode of operation in which the base is common to both the input and output circuits and is usually earthed. The emitter is used as the input terminal and the collector as the output terminal. (Grounded= grounded to AC signals).

■ **common cathode display**

a multisegment light emitting diode (LED) with a single negative voltage input connection. Separate anode connections are provided for each individual segment.

■ **common collector amplifier**

a BJT circuit in which the collector connection is common to both input and output.

■ **common collector connection**

grounded collector connection. Also called the emitter-follower. a mode of operation in which the collector is common to both the input and the output circuits and is usually connected to one of the power rails.

■ **common drain amplifier**

a FET circuit in which the drain connection is common to both input and output.

■ **common emitter amplifier**

a BJT circuit in which the emitter connection is common to both input and output.

■ **common emitter connection**

same as ground emitter connection. a mode of operation for a transistor in which the emitter is common to the input and output circuits. The base is the input terminal and the collector is the output terminal.

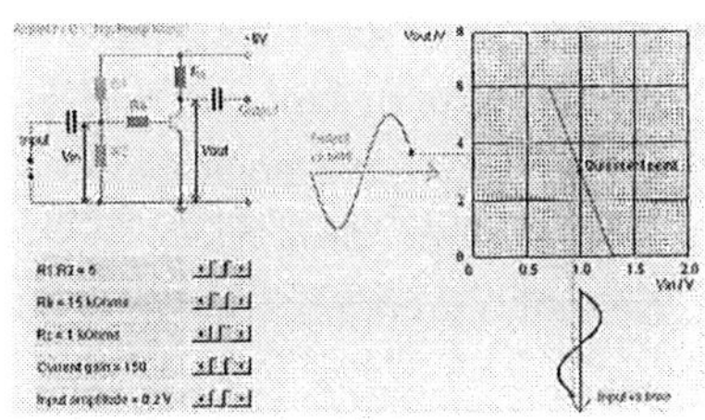

common gate amplifier
a FET circuit in which the gate connection is common to both input and output.

common mode input isolation
on a switching card, the isolation from signal high and low to guard (or shield) for a 3-pole circuit, or from signal high and low to chassis ground for a 2-pole circuit. Specified as resistance and capacitance.

common mode rejection ratio (CMRR)
the ability of an instrument to reject interference from a common voltage at its input terminals with respect to ground. Usually expressed in decibels at a frequency.

common mode voltage
a voltage between input low and chassis ground of an instrument.

common source amplifier
a FET circuit in which the source connection is common to both input and output.

common-anode display
a multisegment light emitting diode (LED) with a single positive voltage input connection. Separate cathode connections are provided for each individual segment.

common-mode rejection (CMR)
1. describes the amount of attenuation the common mode voltage of a differential signal experiences when measured. If the CMR is divided by the system gain (hence input-referred), the term becomes CMRR, or common-mode rejection ratio.
2. the ability of a device to reject a signal that is "common" to or applied to both input terminals. The common-mode signal can be either an AC or a DC signal or a combination of the two. CMR is often expressed in decibels. Common-mode rejection ratio is the ratio of the differential signal gain to the common-mode signal gain.

common-mode signals
signals that appear simultaneously at two inputs of an operational amplifier (op-amp). Common mode signals are always equal in amplitude and phase.

common-mode voltage
describes the amount of voltage that is common to both inputs of a differential input pair.

compact RISC(TM) architecture
an architecture designed by National Semiconductor for reduced instruction set computers.

comparator
1. a particular type of OPERATIONAL AMPLIFIER which has an output that is either on or off. It is a type of switch, but has two inputs, one for what is being measured and the other being a reference point. The comparator will normally switch on when the input is less than the reference. an example would be in a thermostat. Then input would be given by a THERMISTOR and the reference would come from a VARIABLE RESISTOR. This allows the actual switching temperature to be altered.
2.a circuit that compares two voltages a and B (e. g. when a is higher than B then the output goes to a high state). The output is a digital output that is either high or low.

complementary symmetry amplifier
a class B amplifier using matched complementary transistors. Does not require a phase inverter for push-pull output.

complementary transistor
a PNP and NPN pair used in a push-pull circuit.

complementary transistors
two transistors, one NPN and one PNP having near identical characteristics. N-channel and P-channel FETs can also be complementary.

complex numbers
numbers composed of a real number part and an imaginary number part.

compliance
the maximum possible peak-to-peak output of an amplifier.

compliance current
the maximum output current of a constant voltage source. Also known as current limit.

compliance voltage
the maximum output voltage of a constant current source. Also known as voltage limit.

compound semiconductor
a semiconductor material made up of elements from different columns of the periodic table, such as columns III and V or II and VI. III-V semiconductors include gallium arsenide and indium antimonide. II-VI semi-

conductors include zinc sulphide and cadmium sulphide.

■ **conductance**

the degree by which a wire or other component can conduct current. The unit value is the mho, and it is the reciprocal of resistance.

■ **conduction band**

the next higher energy band above the valence band. This represents the energy of charge carriers that have surmounted the energy bandgap and are free to move throughout the material.

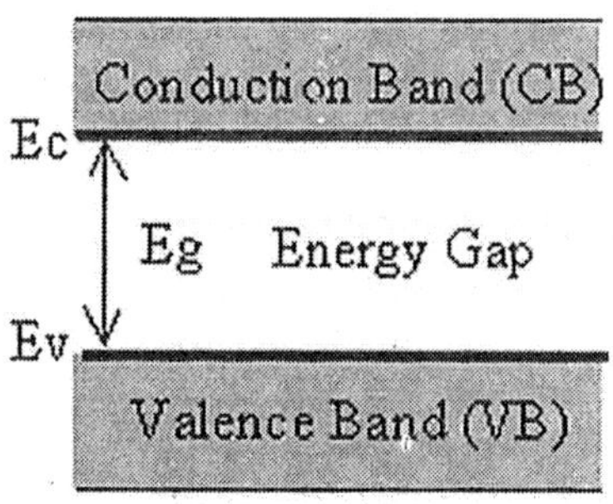

Bands in a Semiconductor

■ **conductivity**

the ability of a wire or component to conduct current.

■ **conductivity type**

n-type or p-type. In n-type extrinsic semiconductor material, electrons are the majority charge carrier. In p-type, extrinsic semiconductor material, holes are the majority charge carrier.

■ **connection path**

the cables, connectors, switch cards, etc. between the device under test (DUT) and the instrumentation. Its major parts are the conductors making the connection and the insulators isolating the conductors from the rest of the world.

■ **connector**

a device that provides for a demountable connection between two fibers or a fiber and an active device and provides protection for the fiber.

■ **constant bandwidth filter**

a band-pass filter whose bandwidth is independent of centre frequency. The filters simulated digitally by the FFT in a DSA are constant bandwidth.

■ **constant current circuit**

circuit used to maintain constant current to a load having resistance that changes.

■ **constant percentage bandwidth**

a band-pass filter whose bandwidth is a constant percentage of centre frequency.1/3 octave filters, including those synthesised in DSAs, are constant percentage bandwidth.

■ **contact**
current carrying part of a switch, relay or connector.

■ **contact bounce**
the intermittent and undesired opening of relay contacts during closure, or closing of relay contacts during opening.

■ **contact life**
the maximum number of expected closures before failure. Life is dependent on the switched voltage, current, and power. Failure is usually when the contact resistance exceeds an end of life value.

■ **contact potential**
1. the voltage generated by the contact of two dissimilar metals or materials.
2.a voltage produced between contact terminals due to the temperature gradient across the relay contacts, and the reed-to-terminal junctions of dissimilar metals. (The temperature gradient is typically caused by the power dissipated by the energized coil.) Also known as contact offset voltage.

■ **contact rating**
the voltage, current, and power capacities of relay contacts under specified environmental conditions.

■ **contact resistance**
for a relay, the resistance in ohms across closed contacts. For a Keithley switching card, also includes the tape resistance and connector terminal resistance.

■ **contact resistances**
resistance per contact. CMOS or bipolar device parameter.

■ **contamination**
generally used to describe the unwanted material that adversely affects the physical, chemical, or electrical characteristics of a semiconductor device.

■ **continuity**
occurs when a complete path for current exists.

■ **continuous wave**
an unmodulated r-f waveform of constant amplitude. Also, a term applied to such a wave which is transmitted in bursts of short and long duration to form the Morse code.

■ **control grid**
the grid (adjacent to the cathode in a tube) to which the signal is applied.

■ **conventional current flow**
concept of current produced by the movement of positive

charges towards the negative terminal of a source.

■ **conventional current flow**
the theory that current flows from a positive terminal and returns to a negative terminal.

■ **convergence**
the bending of light rays toward each other, as by a positive (convex) lens.

■ **conversion rate**
the rate at which sampled analog data is converted to digital data or digital data is converted to analog data.

■ **conversion time**
the time required to complete an analog to digital or digital to analog conversion.

■ **conversion**
a process where a signal is changed from an analog to digital (AD) representation, or digital to analog (DA).

■ **converter**
the mixer stage in a superheterodyne receiver which produces the If signal by heterodyning the r-f carrier with an oscillator signal. Also, a device for changing dc to ac, or ac to dc.

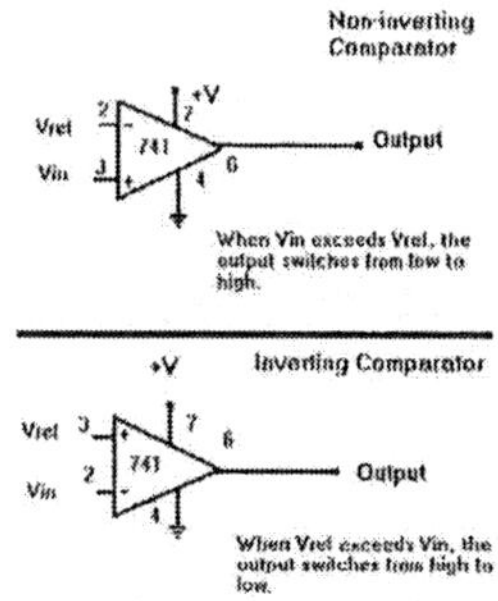

Two Forms of Converter

■ **copper loss**
power lost in transformers, generators, connecting wires and other parts of a circuit due to current flow through the resistance of copper conductors.

■ **core**
1. magnetic material within a coil used to concentrate the magnetic field.
2. the centre of the optical fiber through which light is transmitted.
3. integrated circuit design which is intended to form part of a complete circuit design that incorporates other circuits on the same chip.

■ **corrosion**
an unwanted electrochemical process that affects device (e.g., semiconductor) reliability.

■ **coulomb**
unit of electric charge. a negative coulomb charge consists of 6.23×10^{18} electrons.

■ counter electromotive force
voltage induced into an inductor due to an alternating or pulsating current. Counter emf is always in polarity opposite to that of the applied voltage. Opposing a change of current.

■ counter/timer
a circuit that counts pulses or measures pulse duration.

■ coupler
an optical device that splits or combines light from more than one fiber.

■ coupling
1. to electronically connect two circuits so that signal will pass from one to the other.
2. the linkage established between two circuits or electronic components by use of transformers, capacitors, or inductors.

■ covalent bond
the way some atoms complete their valence shells by sharing valencc electrons with neighbouring atoms.

■ creep
the change in sensor output that occurs with time when a constant measured is applied with environmental and other variable remaining constant. Expressed as a percentage of applied measured over a specified time.

■ crest factor
the ratio of the peak value to the root-mean-square (rms) value of a waveform.

■ critical coupling
that degree of coupling between two resonant circuits which produces a maximum peak of transfer energy at the resonant frequency of the signal.

■ critical machinery
machines which are critical to a major part of the plant process. These machines are usually unspared.

■ critical speed map
a rectangular plot of system natural frequency (y-axis) versus bearing or support stiffness (x-axis).

■ critical speeds
in general, any rotating speed which is associated with high vibration amplitude. Often, the rotor speeds which correspond to natural frequencies of the system.

■ cross axis sensitivity
a measure of off-axis response of velocity and acceleration transducers.

■ cross talk

1. the error output that results from mechanical coupling between the sensing axes on a multi-axis sensor.
2. line distortion caused by wire pairs in the same bundle being used for separate signal transmission.
3. for an analog-to-digital converter (ADC) with more than one input, crosstalk is the amount of signal from one analog input that appears on the measured analog input. This value is typically specified in decibels. For a digital-to-analog converter (DAC), crosstalk is the amount of noise that appears on the DAC output(s) when another DAC is being updated.

■ cross-modulation

the modulation of a desired signal by an undesired signal.

■ cross-neutralisation

the type of neutralisation employed in push-pull r-f amplifiers where the neutralising capacitor is coupled from the anode circuit of one tube to the grid circuit of the other.

■ crossover distortion

distortion caused by both devices in a class B amplifier being cut-off at the same time.

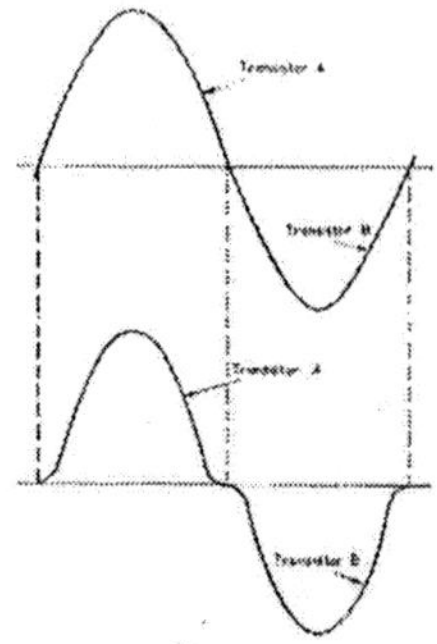

■ crossover frequency

the signal frequency in a multiple-speaker system, where equal signal energy prevails for both the high- and low-frequency systems.

■ cross point

the intersecting point of a column and row in a relay matrix. Specified as (column, row) or (row, column). a cross point generally consists of one or more poles of Form a (normally open) relay switching.

■ crowbar

circuit used to protect the output of a source from a short circuited load. Load current is limited to a value the source can deliver without damage.

■ CRT

abbreviation for cathode ray tube.

■ crystal

natural or synthetic piezoelectric

or semiconductor material with atoms arranged with some degree of geometric regularity.

crystal detector
a solid-state diode demodulator.

crystal oscillator
1. a signal generator in which the frequency is controlled by a piezo-quartz crystal.
2. a circuit whose output is an AC signal (square wave or sine wave) at a fixed frequency (oscillation) set by an external piezoelectric crystal. Changing crystals changes the output frequency. The frequency of a crystal oscillator is more precise than that of an R-C oscillator.

crystal-controlled oscillator
oscillator that uses a quartz crystal in its feedback path to maintain a stable output frequency.

current amplifier
amplifier to increase signal current.

current divider
parallel network designed to divide the total current of a circuit.

current feedback
feedback configuration where a portion of the output current is fed back to the amplifier input.

current feedback amplifier (CFA)
a newer, high performance architecture that delivers higher bandwidth and slew rates at much lower supply currents.

current gains
a bipolar device parameter. Ratio of collector to base (beta) or collector to emitter (alpha) currents.

current mirror
term used to describe the fact that DC current through the base circuit of a class B amplifier is approximately equal to the DC collector current.

current saturation
maximum flow of electric force in a conductor; in a laser, the point at which further electrical charge will not increase lasing action.

current-limiting resistor
resistor in the path of current flow to control the amount of current drawn by a device.

cutback method
a technique for measuring the loss of bare fiber by measuring the optical power transmitted through a long length then cutting back to the source and mea-

suring the initial coupled power.

■ **cut-off**

condition when an active device is biased such that output current is near zero or beyond zero.

■ **cut-off frequency**

frequency at which the power gain of an amplifier falls below 50% of maximum.

■ **cut-off wavelength**

the wavelength beyond which single mode fiber only supports one mode of propagation.

■ **cycle**

when a repeating wave rises from zero to a positive maximum then back to zero and on to a negative maximum and back to zero it is said to have completed one cycle.

■ **D/A (DIGITAL-TO-ANALOG) CONVERTER**

an electronic device, often an integrated circuit, that converts a digital value to an analog voltage. D/A converters are used in many instruments to convert digital reading information into an analog signal for analog output.

■ **D/A**

abbreviation for digital-to-analog.

■ **damping**

1. reduction in magnitude of oscillation due to energy being dissipated as heat.
2. materials, design, and mounting techniques used to reduce ringing in the transducer.
3. the quality of a mechanical system that restrains the amplitude of motion with each successive cycle. Damping of shaft motion is provided by oil in bearings, seals, etc.The damping process converts mechanical energy to other forms, usually heat.

■ **Darlington pair**

an amplifier consisting of two bipolar junction transistors with their collectors connected together and the emitter of one connected to the base of the other. Circuit has an extremely high current gain and input impedance.

■ **data transfer**

refers to the way data is transferred to and from memory, such as programmed I/O or DMa mode.

■ **dB (Decibels)**

a technique for expressing voltage, power, gain, loss, or frequency in logarithmic form against a reference. Typical ref-

erences include volts, watts or Hz. Decibels are calculated using the expression
dB = 10*log(x/y).

■ dBd
decibels over dipole, a relative gain measurement with respect to a half wave dipole (0 dBd – 2.14 dBi) using a standard dipole antenna as a reference.

■ dBi
decibels isotropic, a relative gain measurement with respect to an isotropic radiator in free space (uniform emitter in free space, a theoretical situation). Normally used to describe gain for antennas operating at 1GHz or above.

■ dBm
optical power referenced to 1 milliwatt.

■ DC
direct Current; the external, constant-voltage, single polarity electrical power supplied to an integrated circuit.

■ D-C amplifier
an amplifier employing direct coupling.

■ DC continuity
the ability of a system to return a DC current. Some antennas are a DC open (any monopole type antenna), some DC short (any folded dipole antenna). Please note that DC continuity does not directly relate to RF continuity.

■ DC ground
a low impedance conductor, capable of carrying away unwanted RF interference/static/noise

■ DC load line
a graph representing all possible combinations of voltage and current for a given load resistor in an amplifier.

■ DC offset
the change in input voltage required to produce a zero output voltage when no signal is applied to an amplifier.

■ DC parametric
the operating characteristics of an integrated circuit or discrete device that can be measured with the device in a static condition.

■ DC power supply
any source of DC power for electrical equipment.

■ DC Restorer
a damper circuit which restores the d-c level to a signal waveform.

■ dead short
short circuit having zero resistance.

■ decade
a frequency factor of ten.

■ decibels (dB)
a logarithmic representation of amplitude ratio, defined as 10 times the base ten logarithm of the ratio of the measured power to a reference. dB V readings, for example, are referenced to 1 volt rms. dB amplitude scales are required to display the full dynamic range of a DSA. dB values for power or voltage measurements yields the same result.

■ decoupling circuit
a resistive-capacitive circuit designed to isolate the signal-carrying circuits with respect to common coupling through the power-supply system.

■ de-emphasis circuit
an RC filter section employed after FM detector systems to decrease high-frequency signal levels which had been increased during transmission.

■ deep depletion
the phenomenon of driving the space charge or depletion region of the semiconductor material deeper than in equilibrium by sweeping or stepping the voltage quickly in the direction from accumulation toward inversion.

■ deep level transient spectroscopy (DLTS)
an advanced technique for studying the interface *dynamic*, not just density. Along with DIT, DLTS measures trap capture cross-section.

■ defect
an imperfection in the surface structure of a silicon wafer which causes one or more chips to malfunction.

■ degeneration
the attenuation of a signal due to inverse feedback.

■ degenerative feedback
also called negative feedback. a portion of the output of an amplifier is inverted and connected back to the input. This controls the gain of the amplifier and reduces distortion and noise.

■ degrees of freedom
a phrase used in mechanical vibration to describe the complexity of the system. The number of degrees of freedom is the number of independent variables describing the state of a vibrating system.

■ de-ionised time
the time interval for ionised gas

in a tube to reach the de-ionized condition.

delay line
any ladder filter or transmission line which introduces a delay in time for a signal passing through.

delay spread
a type of distortion due to multipath resulting in the spreading out or "smearing" of the received signal. It occurs when identical signals arrive via different paths and have different time delays.

delay time
the time for collector current to reach 10% of its maximum value in a BJT switching circuit.

delayed AVC
an automatic volume-control circuit which is designed to produce an AVC bias only for signals above a predetermined amplitude.

delta-sigma modulators
circuits which convert analog signals into a bit stream (a circuit which uses a specific technique for converting analog signals to digital signals), enabling the bit stream to be processed in the digital domain.

demodulation
a signal-rectifying system which derives the modulating-signal component from the modulated carrier.

depletion approximation
a type of doping profile calculation from a C V that only takes into account the theory of depletion operation (not accumulation or inversion). Because of this approximation, it is not accurate at very small or great depths..

depletion layer
the region of a semiconductor device where essentially all charge carriers are swept out by an electric field.

depletion mode
in a FET, an operating mode where reverse gate-source voltage is used to deplete the channel of free carriers. This reduces the sise of the channel and increases its resistance.

depletion region
the area surrounding a pn junction that is depleted of carriers.

depletion-mode MOSFET
a MOSFET designed to operate in either depletion mode or enhancement mode.

■ **deposition**

the procedure in which materials are deposited onto a substrate by means of vacuum, electrical, chemical, screening or vapour techniques.

■ **depth of field**

the working range of the beam, a function of wavelength, diameter of the unfocused beam, and focal length of the lens. To achieve a small diameter spot size, and thus a high power density, a short depth of field must be accepted.

■ **depth**

the distance between the semi-conductor-insulator interface and the depletion layer edge.

■ **design kit**

a collection of technical information, software, and computer data files which enable the designer to model, simulate, and layout integrated circuits using a specific technology (e. g. 0.8 micron CMOS).

■ **detector**

a photodiode that converts optical signals to electrical signals.

■ **deviation ratio**

the ratio of maximum FM carrier deviation to the highest-frequency audio-modulating signal employed.

■ **diac**

a two terminal bidirectional thyristor. Has a symmetrical switching mode.

■ **dialectric constant**

property of a material that determines how much electrostatic energy can be stored per unit volume when unit voltage is applied.

■ **dialectric strength**

the maximum voltage an insulating material can withstand without breaking down.

■ **dielectric permitivity**

the constant of proportionality between a capacitor's geometry and its capacitance.
capacitance = (dielectric permitivity) * area / dielectric thickness.

■ **dielectric**

an insulating layer. a material that has high resistance. This term is usually used when the insulating layer separates the plates of a capacitor.

■ **differential ADC**

an A/D converter with differential inputs. This means that the input signal is represented by the difference between two voltages, thus greatly reducing noise and other disturbing factors.

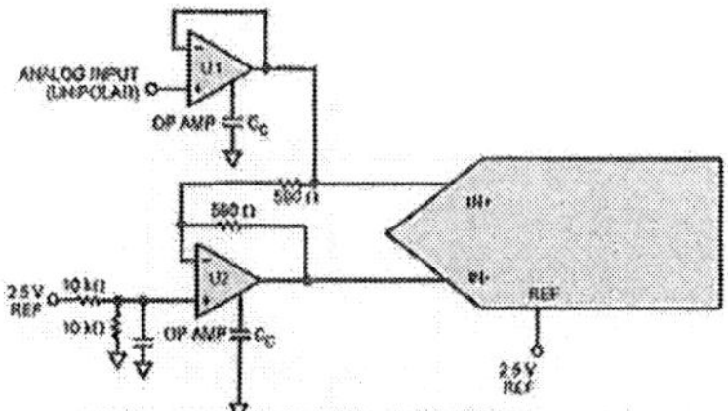

A Single-ended 16-Bit ADC Driver Circuit

■ **differential amplifier**

1. two transistors connected as a differential pair, with a third transistor (tail transistor) connected to the common emitters

2 .an op amp configured to amplify the difference between 2 input signals.

3. an amplifier in which the output is in proportion to the differences between voltages applied to its two inputs.

■ **differential amplifier**

a particular type of OPERATIONAL AMPLIFIER where the output voltage is in proportion to the difference between it's inputs. Unlike the COMPARATOR, which can based on exactly the same INTEGRATED CIRCUIT, the Op-Amp has two individual inputs instead of one input and one reference value. This use is sometimes called a PROPORTIONAL AMPLIFIER or SUBTRACTOR.

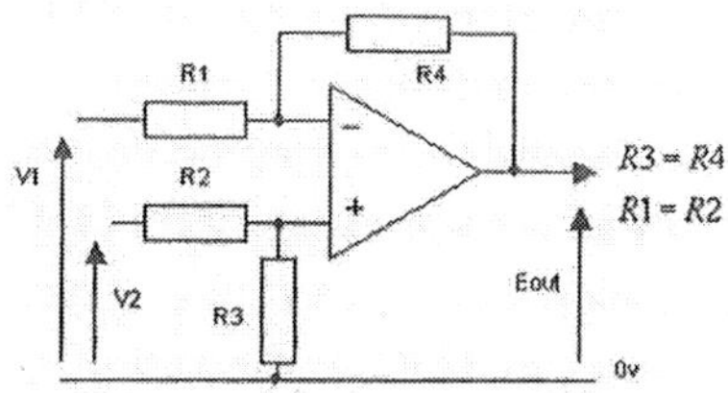

■ **differential input isolation**

on a switching card, the isolation from signal high to low. Specified as resistance and capacitance.

■ **differential inputs**

signals that are represented by the difference between two voltages or currents are referred to as differential signals. In effect, differential inputs subtract two input signals from each other. The result is lower noise as noise present on both inputs is subtracted away, leaving just the signal.

■ **differential nonlinearity (DNL) error**

assuming an ideal analog-to-digital converter (ADC) with finite digital codes exactly 1LSB apart (DNL error = 0) or an ideal digital-to-analog converter (DAC) with analog output values exactly one code apart (DNL error = 0), the DNL error is defined as the difference between the ideal and the measured code transitions for successive codes for an ADC or the difference between the ideal and the measured output value between successive DAC codes. As an example, when using an ADC, as the analog input voltage increases, if the next code transition occurs at a voltage 1. 5LSB away from the previous transition, the DNL error is 0.5LSB; if it occurs at 0.5LSB away, the DNL error is -0.5LSB.

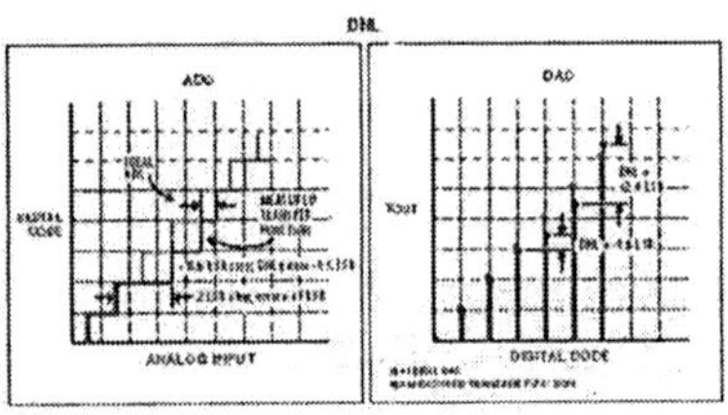

■ **differential non-linearity**
the maximum deviation of an actual quantised step width from the ideal quantised step width.

■ **differential pair**
two transistors connected in a differential configuration .

■ **differentiating circuit**
a short time-constant RC or L/R circuit used for pulse or square-wave signal modification.

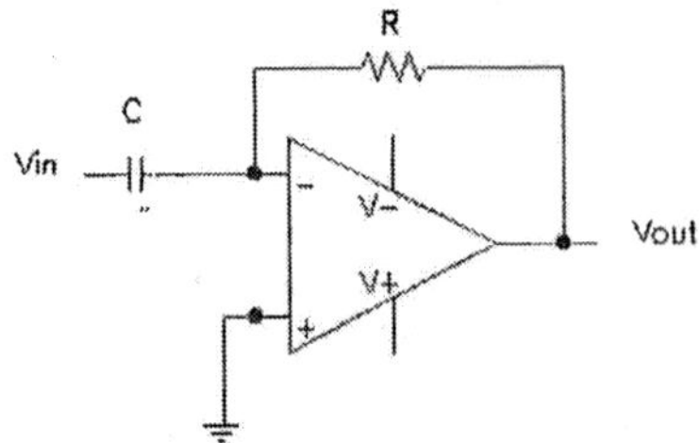

■ **differentiation**
representation in terms of time rate of change. For example, differentiating velocity yields acceleration. In a DSA, differentiation is performed by multiplication by jω in the frequency domain, where ù is frequency multiplied by 2ω. (Differentiation can also be used to convert displacement to velocity.).

■ **differentiator**
a circuit in which the output voltage is in proportion to the rate of change of the input voltage. a high pass RC circuit.

■ **diffraction**
1. a type of distortion due to multipath resulting in the spreading out or "smearing" of the received signal. It occurs when identical signals arrive via different paths and have different time delays.
2. a propagation phenomenon that allows radio waves to

propagate beyond obstructions via secondary waves created by the obstruction. Classic types of diffractions are smooth earth and knife-edge.

■ **diffusion**
tendency of conduction band electrons to wander across a pn junction to combine with valence band holes.

■ **diffusion method**
diffusion methods place a sensor at the point of measurement and expose it directly to a gas.

■ **digital**
relating to devices or circuits that have outputs of only two discrete levels. Examples: 0 or 1, high or low, on or off, true or false etc.

■ **digital circuit**
an electronic circuit in which the intensity of an external quantity (e.g., sound level) is represented by a series of fixed voltages using a numerical code (e.g., the binary code requires only two voltage levels).

■ **digital electronics**
the branch of electronics dealing with information in binary form.

■ **digital feed through**
the amount of noise that appears on the digital-to-analog converter (DAC) output when the DAC digital control lines are toggled.

■ **digital filter**
a filter which acts on the data after it has been sampled and digitised. Often used in DSAs to provide anti-aliasing protection before internal re-sampling.

■ **digital I/O**
abbreviation for digital input/output.

■ **digital lines/ports/bits/channels**
in hardware, a digital line is physical hardware connection to a pin with a digital signal. a digital port is a physical grouping of digital lines. In software, a digital bit (1 or 0) is a logical representation of a digital line. a digital channel is a logical grouping of digital bits.

■ **digital signal**
a signal that varies only at regular time intervals and has one, two, or more predetermined amplitudes for each interval.

■ **digital trigger**
an event that occurs at a user-selected point on a digital input signal. The polarity and sensitivity of the digital trigger can often be

programmed.

■ **diode**

1. a semiconductor that allows current to flow in one direction only.

2. a two-element tube or two-terminal solid-state rectifier.

3. a diode will allow electrical current to flow in one direction only. They are SOLID STATE devices and typically need more than 0.6 VOLTS to operate. In a normal diode, current flows from the positive ANODE to the negative CATHODE. There are several special types including LEDs, PHOTODIODES and ZENER DIODES. Diodes only allow current to flow in one direction. They act like a one way valve. They are passive devices.

■ **diode array**

multiple discrete (usually unconnected) diodes on a single silicon chip.

■ **diode breakdown voltage**

the voltage under specified conditions at which there is a sudden change from high dynamic resistance to a much lower value.

■ **diode drop**

the forward voltage developed across a diode when it is operating.

■ **DIP**

abbreviation for "dual in line package."

■ **dipole**

1. a balanced antenna, usually half a wavelength long and fed (or taken off) at its centre.

2. a monopole with a ground balance (mirror image).

■ **dipole array**

a group of dipoles use to transmit/receive together as one.

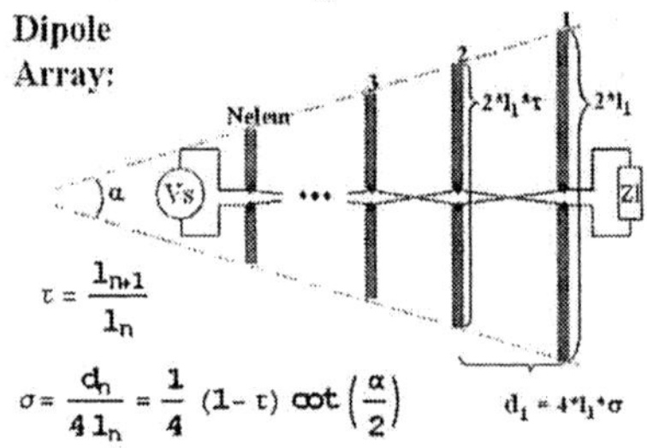

An Example-Log Periodic Dipole Array, LPDA

■ **direct coupled amplifier**

an amplifier in which the output of one stage is coupled to the input of the next without the use of a capacitor. This type of amplifier will amplify Direct Currents and low frequency waveforms.

■ **direct coupling**

where the output of an amplifier is connected directly to the input of another amplifier or to a load. Also known as DC coupling because DC signals are not blocked.

direct current
Current that flows in only one direction.

direct digital synthesis
a technique for signal generation where the signal is directly synthesised using only digital techniques. This technique generates very precise waveforms, even at low frequencies. Waveforms with correct phase and frequency are obtained immediately after a shift to a new frequency.

direction discriminator:
part of a bi-directional counter which determines the counting direction.

directivity
the property of radiating more strongly in a particular direction.

directivity Index
the value in dB of ten times the common logarithm of the directivity factor. The directivity factor is the ratio of the sound intensity produced by a test transducer on a specific axis to that of a point source that is putting out the same acoustic power. Since the specific axis is usually one of maximum radiation, the DI in usually greater than zero.

discharge
release of energy stored in either a battery or a capacitor.

discharge circuit
the term is usually applied to a circuit which produces a sawtooth waveform.

discrete
a term used for circuit made up of individual components.

discrete device
a class of electronic components that contain one active element, such as a transistor or diode. However, hybrids, optoelectronic devices, and intelligent discrete may contain more than one active element.

discrete fourier transform
a procedure for calculating discrete frequency components (filters or lines) from sampled time data. Since the frequency domain result is complex (i.e., real and imaginary components), the number of frequency points is equal to half the number of time samples (for a real FFT). When using zoom analysis, the FFT uses complex time data and then the number of frequency lines is

equal to the number of time samples.

discriminator

a circuit for demodulating frequency modulation or for producing an error signal when comparing two a-c signals.

dispersion

the temporal spreading of a pulse in an optical wave guide. May be caused by modal or chromatic effects.

displacement

the change in distance or position of an object relative to a reference.

displacement current

the current that flows through a capacitor in response to the rate of voltage change across it: displacement current = C * dV/dt.

displacement transducer

a transducer whose output is proportional to the distance between it and the measured object(usually the shaft).

dissipation

release of electrical energy in the form of heat.

distance-coded reference marks

incremental measuring method, whereby the absolute position can be
determined by evaluating the systematically varying distances between
consecutive reference marks.

distortion

1. an undesired change in a waveform or signal.
2. the amount by which a circuit or component fails to reproduce accurately the characteristics of the input.

distributed capacitance

any capacitance other than that within a capacitor. For example, the capacitance between adjacent turns of wire in a coil.

distributed capacitance

the term generally applied to the capacitance set up between turns and layers of an inductor.

distributed inductance

any inductance other than that within an inductor. Example inductance in any conductor.

divergence

the angle at which the laser beam spreads in the far field; the bending of rays away from each other, as by a concave lens or convex mirror.

■ **DMA (DIRECT MEMORY ACCESS) channels**

is a bus PCs offer eight parallel channels for DMA mode data transfers. a number of these are reserved for exclusive use by the computer. The remainder are available for use by user-supplied I/O options, such as plug-in data acquisition cards. Also called DMA levels.

■ **DMA mode**

mode in which data transfers directly between an I/O device and computer memory. The CPU is bypassed..

■ **DMM**

an electronic instrument that measures voltage, current, resistance, or other electrical parameters by converting the analog signal to digital information and display. The typical five-function DMM measures DC volts, DC amps, AC volts, AC amps, and resistance.

■ **domain**

a moveable magnetized area in a magnetized material. Also known as magnetic domain.

■ **donor atoms**

pentavalent atoms that give up electrons to the conduction band in an N type semiconductor material.

■ **donor**

a material added as a dopant to a semiconductor to make it n-type by "donating" valence electrons, which can conduct electric charge. Phosphorus is a donor dopant for silicon.

■ **dopant**

a material added in minute quantities (about 1ppm) to a semiconductor to alter its electrical conductivity type. Dopants may be donors or acceptors.

■ **doping**

the process of adding impurity atoms to intrinsic (pure) silicon or germanium to improve the conductivity of the semiconductor material.

■ **doping profile**

the profile or distribution of dopants in the semiconductor material. In other words, the graph of dopant concentration at each depth into the semiconductor device.

■ **dot convention**

standard used with transformer symbols to indicate whether the secondary voltage is in phase or out of phase with the primary voltage.

■ **doubler**

a circuit in transmitting systems which doubles the frequency of the input signal. Also, a circuit in power-supply systems for doubling voltage amplitude.

■ **drain terminal**

along with the source and gate, one of the three terminals of a field effect transistor (FET).

■ **DRAM**

dynamic Random Access Memory; memory circuits which require regular refreshing of the data stored in each memory cell. The circuits are more compact than static RAM but do not retain data when the refresh signal (clock) or power is turned off.

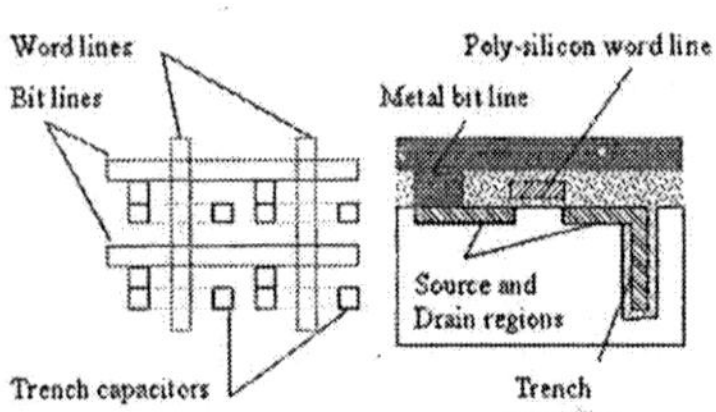

(a) (b)

■ **DRC**

Design Rules Check; operation carried out by a CAD tool which compares the dimensions, separations, overlaps, etc. of a chip layout to the minimum and maximum dimensions specified by the design rules for a particular process.

■ **drift**

a problem that can develop in tuned amplifiers when the frequency of the tuned circuit changes due to temperature or component aging.

■ **drift, angular**

all undesirable variations in output (either amplitude or frequency); angular drift of the beam, measured in milliradians before, during, and after warm-up.

■ **driver stage**

an audio or r-f amplifier stage preceding the final high-power amplifier.

■ **drivers**

software that controls a specific hardware device such as a data acquisition board.

■ **driving point cable correction**

a method of correcting for errors in high frequency capacitance measurements due to cables connecting the DUT to the instrument. In this method, the cable impedance is measured by the instrument with the DUT end open.

■ **dropping resistor**
resistor whose value has been chosen to drop or develop a given voltage.

■ **dry circuit switching**
switching below specified levels of voltage (e.g., 20mV) and current to minimise any physical and electrical changes in the contact junction.

■ **dry reed relay**
a glass-enclosed, hermetically sealed, magnetically actuated contact. No mercury or other wetting material is used.

■ **DSP**
digital Signal Processing; the manipulation of information which has been numerically coded using a digital circuit. Analog data (signals) are first converted to digital form by an analog-to-digital (A/D) converter, then processed using digital circuits, then converted back to analog form by a digital to analog (D/A) converter.

■ **dual trace oscilloscope**
oscilloscope that can simultaneously display two signals.

■ **duty cycle**
the length of time the laser beam is actually cutting, drilling, welding, or heat-treating, as compared to the entire work cycle time.

■ **duty ratio**
the ratio of pulse width to repetition period. Also known as Duty Cycle.

■ **dynamic**
relating to conditions that are changing or in motion.

■ **dynamic signal analyser (DSA)**
vibration analyser that uses digital signal processing and the Fast Fourier Transform to display vibration frequency components. DSAs also display the time domain and phase spectrum, and can usually be interfaced toa computer.

■ **dyne**
the unit of force in the cgs (centimeter-grams-second) system equal to that required to produce, in a 1-gram mass, an acceleration of 1 centimeter per second for every second that the force is applied.

■ **earphone**
a device for tuning electrical energy into sound waves - fits in your ear.

earth

this is another name for the GROUND, or zero VOLT connection of a circuit. Sometimes called neutral, though this can cause confusion with NEGATIVE supply.

e-beam

electron beam; a focussed beam of high energy electrons used to expose photo-sensitive film in the production of integrated circuit masks (photo-masks), or the circuit elements themselves.

eccentricity error

measuring error of an angle encoder caused by an eccentricity in the
mounting of the circular graduation.

eccentricity ratio

the vector difference between the bearing centre line and the average steady-state journal centre line.

echo location

determining the location of a target relative to the sensor face by means of measuring the time it takes for a sound wave to travel to the target and be reflected back to the sensor.

ECL Emitter Coupled Logic

Where transistors are held in the turned-on state to increase the speed of the gate.

ECL

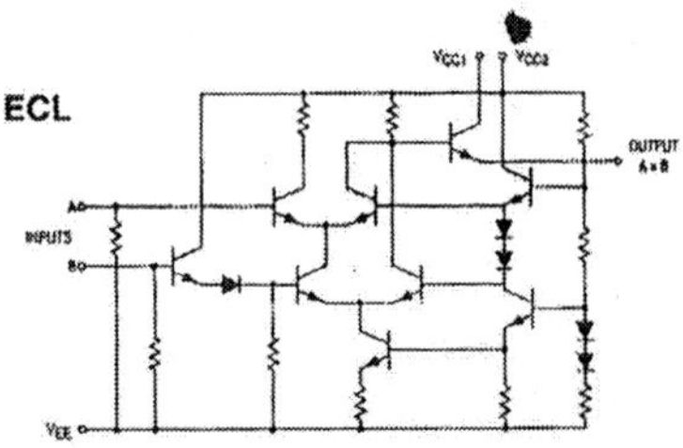

EDA

Electronic Design Automation; the name used for activities or facilities that involve software design aids used in chip design, or the industry sector making the aids.

eddy current

electrical current which is generated (and dissipated)in a conductive material in the presence of an electromagnetic field.

EDFA

erbium-doped fiber amplifier, an all optical amplifier for 1550 nm SM transmission systems.

edge-emitting diode (E-LED)

a LED that emits from the edge of the semiconductor chip, producing higher power and narrower spectral width.

■ EEPROM or EPROM.
Electrically-Erasable Programmable Read-Only Memory. Similar to PROM, but with the capability of selective erasure of information through special electrical stimulus. Information stored in EEPROM chips is retained when the power is turned off.

■ effective base width (parametric testing)
a bipolar device parameter. Base width at specified collector/base voltage.

■ effective channel lengths, linear. (parametric testing)
a CMOS device parameter. Electrical channel length from scaling devices, small drain voltage.

■ effective channel lengths, saturation (parametric testing)
a CMOS device parameter. Electrical channel length from self-consistency, large drain voltage.

■ effective channel width. (parametric testing)
electrical channel width from scaling devices, small drain voltage.

■ effective number of bits (ENOB)
the measured performance (in bits) of an analog-to-digital converter (ADC) with respect to input frequency f_{IN}. As f_{IN} increases, overall noise (particularly the distortion components) also increases, thereby reducing the ENOB and SINAD.

■ effective value
the value of alternating current or voltage equal to the product of 0.707 times the peak amplitude. It is the electric-energy equivalent of the same value of direct current.

■ efficiency
the amount of power delivered to the load of an amplifier as a percentage of the power required from the power supply.

■ efficiency
in a projector efficiency is defined as the ratio of the acoustic power generated to the total electrical power input. Efficiency varies with frequency and is expressed as a percentage.

■ EISA bus
extended Industry Standard Architecture. a 32-bit wide, upwards-compatible extension of the 16-bit wide ISa bus. Sometimes called the XT bus.

electric charge
electric energy stored on the surface of a material. Also known as a static charge.

electric field
a field or force that exists in the space between two different potentials or voltages. Also known as an electrostatic field.

electric polarisation
a displacement of bound charges in a dielectric when placed in an electric field.

electric vector
the electric field associated with a light wave and having both direction and amplitude. Commonly represented by a line with an arrowhead.

electrical runout
an error signal that occurs in eddy current displacement measurements when shaft surface conductivity varies.

electroacoustic transducer
device that produces an energy transfer from electric to acoustic (sound) or from acoustic to electric. Examples include a microphone, earphones and loudspeakers.

electrode
an electrical terminal, normally left unconnected at one end. Electrical connection between electrodes is often made by electric arcing or holding them in salt solutions.

electroluminescence
conversion of electrical energy into light energy.

electrolyte
electrically conducting liquid (wet) or paste (dry).

electrolytic
a type of capacitor that has a liquid or paste between the plates to increase its capacitance.

electrolytic capacitor
a capacitor having an electrolyte between the two plates. a thin layer of oxide is deposited on only the positive plate. The oxide acts as the dielectric for the capacitor. Electrolytic capacitors are polarized and so must be connected in correct polarity to prevent breakdown.

electromagnetic communication
use of an electromagnetic wave to pass information between two points. Also called wireless communication.

electromagnetic wave
wave that consists of both electric and magnetic variation.

electromechanical transducer
device that transforms electrical energy into mechanical energy (electric motor) or mechanical energy into electrical energy (generator).

electrometer
a highly refined DC multimeter. In comparison with a digital multimeter, an electrometer is characterized by higher input resistance and greater sensitivity. It can also have functions not generally available on DMMs (e.g., measuring electrical charge, sourcing voltage).

electromotive force
force that causes the motion of electrons due to potential difference between two points. (voltage).

electron
smallest sub atomic particle of negative charge that orbits the nucleus of an atom.

electron affinity
the energy required to move an electron from the bottom of the semiconductor conduction band to the vacuum energy level, i.e. to infinity. This number is used in the calculation of work function in devices with polysilicon gates.

electron flow
the direction in which electrons flow. This is from negative to positive - as electrons are negatively charged.

electron gun
the term generally applied to the tube electrodes in a cathode-ray tube which produce a narrow electron beam.

electron-coupled oscillator
an oscillator which isolates the load circuit from the oscillator section by coupling the two by the electron stream within the tube.

electron-hole pair
the electron and hole that are simultaneously generated thermally or optically. Once generated, the pair can either recombine and disappear or separate and conduct charge as independent free carriers.

embedded applications
applications which have been integrated with other functions on a single integrated circuit.

emitter
the semiconductor region from which charge carriers are injected into the base of a bipolar junction transistor.

emitter feedback
coupling from the emitter output to the base input of a bipolar junction transistor.

emitter follower
a common collector amplifier. Has a high current gain, high input impedance and low output impedance.

encoder
apparatus consisting of a measuring standard and a scanning unit (transducer, sensor).

end finish
the quality of the end surface of a fiber prepared for splicing or terminated in a connector.

energy band
the specific levels of energy that charge carriers (electrons or holes) in a solid are permitted to have. Carriers cannot have an energy outside the allowed energy band.

engineering notation
a floating point system in which numbers are expressed as products consisting of a number greater than one multiplied by an appropriate power of ten that is some multiple of three.

enhanced pulsing
electronic modulation of a laser beam to produce high peak power at the initial stage of the pulse. This allows rapid vaporization of the material without heating the surrounding area. Such pulses are many times the peak power of the CW mode.

enhancement-mode MOSFET
a field effect transistor in which there are no charge carriers in the channel when the gate source voltage is zero.

epitaxial layer
a deposited layer of material having the same crystallographic characteristics as the substrate material.

epitaxy
the process of growing a single crystal film onto another crystalline substrate or layer. The resulting "epitaxial layer" is sometimes referred to as "epi."

EPROM
Erasable Programmable Read Only Memory; read only memory (see below) which can be preprogrammed by the user

and subsequently erased and preprogrammed with new data as the need arises.

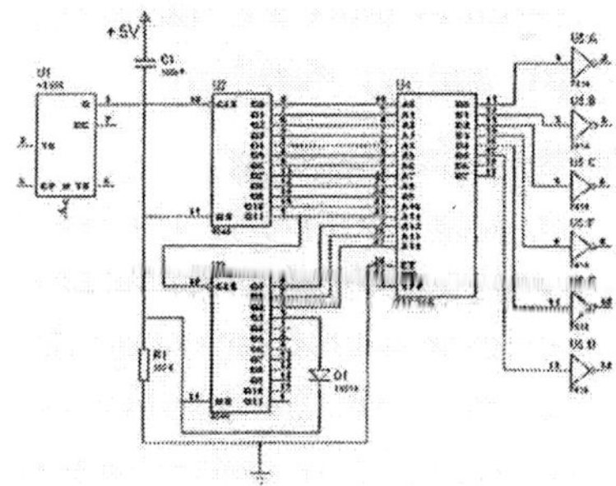

Erasable Programmable Read Only Memory

■ equilibrium modal distribution (EMD)

steady state modal distribution in multimode fiber, achieved some distance from the source, where the relative power in the modes becomes stable with increasing distance.

■ equilibrium

the condition in a semiconductor device or material when there is no longer any tendency for its macroscopic properties to change with time. In other words, everything that is going to happen has happened. Specifi cally, in an MOS capacitor, current ceases to flow and the capacitor is fully charged.

■ equivalent resistance

total resistance of all the individual resistances in a circuit.

■ error amplifier

an amplifier placed in a feedback loop to improve system response.

■ error

the deviation (difference or ratio) of a measurement from its true value.

■ ESCON

IBM standard for connecting peripherals to a computer over fiber optics. Acronym for Enterprise System Connection.

■ etch

a chemical process used to selectively remove parts of an integrated circuit during fabrication.

■ excess loss

the amount of light lost in a coupler, beyond that inherent in the splitting to multiple output fibers.

■ exclusive or Logic

element that features two inputs. The output will be 1 only when one or the other (but not both) is logic . 1. In all other cases the output is 0.

■ external sampling

in a DSA, refers to control of data sampling by a multiplied tachometer signal. Provides a stationary display of rpm-related peaks with changing speed.

external trigger

an analog or digital hardware event from an external source that starts an operation..

extrinsic base resistances (parametric testing)

a bipolar device parameter. Base resistance outside active base.

extrinsic channel resistances (parametric testing)

a CMOS device parameter. Sum of all resistances between channel and source/drain connections.

extrinsic semiconductor

a semiconductor that has been doped either n-type or p-type. Electrons and holes are present in unequal proportions (by 4 to 8 orders of magnitude).

FAB.

in semiconductor manufacturing, fabrication usually refers to the front-end process of making devices and integrated circuits in semiconductor wafers, but does not include the package assembly (back-end) stages.

fall time

time it takes the falling edge of a pulse to go from 90% of peak voltage to 10% of peak voltage.

farad

the unit of capacitance. "F" This is a very large value and the most common units are microfarad (μF) and nanofarad (n).

far-field imaging

an imaging technique with solid-state lasers that has several limitations non-uniform energy distribution, very short working distances, and poor control of hole geometry.

Fast Fourier Transform (FFT)

a computer (or microprocessor) procedure for calculating discrete frequency components from sampled time data. a special case of the Discrete Fourier Transform, DFI, where the number of samples is constrained to a power of 2 for speed.

feed element

a dipole arrangement that can be inserted into a larger array of elements (a grid array for example) which increase the performance of the feed element Front to Back Ratio (FTB) The difference (logarithmic subtraction) between the antenna's maximum gain (normally in the direction the antenna is pointed) and the gain in the opposite direction (i.e.

10dBd-(-12dBd) gives a FTB of 22dB). It does not matter if you use dBi or dBd as long as they are consistent).

■ **feedback**

a portion of the output signal of an amplifier which is connected back to the input of the same amplifier.

■ **feedback amplifier**

an amplifier with an external signal path from its output back to its input.

■ **feedback oscillator**

a signal-generating circuit which employs regenerative feedback to sustain oscillations.

■ **fermi level**

the energy level in the semiconductor device at which the probability of finding an electron is 50%. In other words, it's like the water level in a glass. States below the Fermi level are full and those above it are empty.

■ **ferrite rod aerial**

a coil of wire wound on a ferrite material to increase the inductance of the coil. It's signal capturing capability.

■ **ferrite-core inductor**

an inductor wound on a ferrite core.

■ **ferrule**

a precision tube which holds a fiber for alignment for interconnection or termination. a ferrule may be part of a connector or mechanical splice.

■ **fiber amplifier**

an all optical amplifier using erbium or other doped fibers and pump lasers to increase signal output power without electronic conversion.

■ **fiber identifier**

a device that clamps onto a fiber and couples light from the fiber by bending, to identify the fiber and detect high speed traffic of an operating link or a 2 kHz tone injected by a test source.

■ **fiber optics**

laser's light output carries information that is conveyed between two points by thin glass optical fibers.

■ **fiber tracer**

an instrument that couples visible light into the fiber to allow visual checking of continuity and tracing for correct connections.

■ **fidelity**

the quality of reproducing an output signal exactly like the input signal except for amplitude (and somtimes phase); i.e., output and input signals exactly alike in terms of frequency and shape.

■ **Field Effect Transistor (FET)**
a transistor that makes use of the field established in a p-type channel semiconductor material to control the flow of current through the channel.

■ **field oxide threshold voltage (parametric testing)**
a CMOS device parameter. Voltage at 40nA*W/L or predefined current, thick-oxide MOSFETs.

■ **FIFO**
first-in/first-out memory buffer. The first data into the buffer is the first data out of the buffer.

■ **filament**
that electrode within a tube which is heated for electron emission or which transfers its heat to a separate cathode for electron emission.

■ **filter**
1. a circuit that performs filtering, i.e., it removes some undesired signals from a desired signal. Typically, a filter removes (attenuates) signals in a particular frequency range.
2 network consisting of capacitors, resistors and/or inductors used to pass certain frequencies and block others.

■ **filter circuit**
a circuit designed to filter certain signal components and to prevent their transfer to a subsequent circuit or device.

■ **filtering**
the task performed by electrical circuits which pass frequencies within a specified frequency band and attenuate signals that fall outside that frequency band.

■ **fixed oxide charge**
charge trapped in the insulator of an MIS device that cannot move to the semiconductor material. Unlike oxide trapped charge, fixed oxide charge does not anneal out during a BTS cycle. This charge may be located at either oxide interface and is commonly produced by carrier injection or radiation..

■ **flash memory**
a non-volatile storage device whereby erasing data can only be done in blocks or in relation to the entire chip.

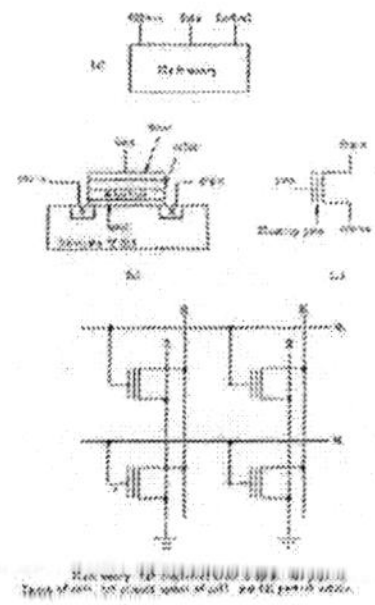

flash lamp

source of powerful light; often in the form of a helical coil and used to excite photon emission in a solid-state laser.

flat top filter

FFT window function which provides the best amplitude accuracy for measuring discrete frequency components. Note there are several different flat top windows. The HP proprietaryP401 is the "best" flat top window, P301 is the most common.

flip flop

an astable multivibrator. a square wave oscillator that has no stable states. Also half a shift register. Flip Flops can be unclocked and triggered by the input pulses or CLOCKED by a clock pulse to a special clock input.

flow soldering

flow or wave soldering technique in large scale electronic assembly to solder all the connections on a printed circuit board by moving the board over a wave of molten solder.

fluid-film bearing

a bearing which supports the shaft on a thin film of oil. The fluid-film layer may be generated by journal rotation (hydrodynamic bearing), or by externally applied pressure(hydrostatic bearing).

fluorescence

the glow induced in a material when bombarded by light. Brewster windows of fused silica fluoresce in UV light, increasing absorption of laser radiation and degrading laser mode and output.

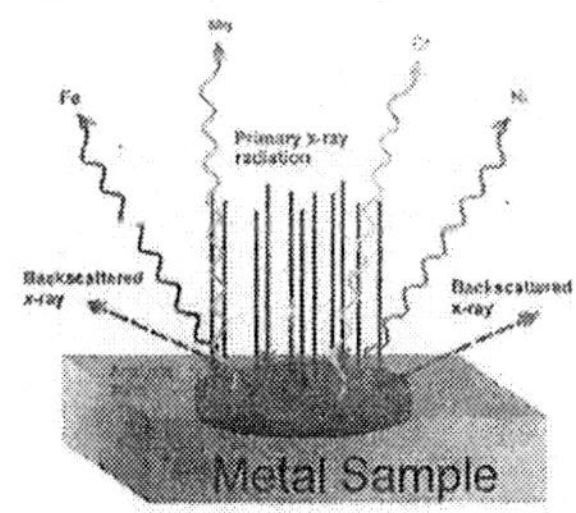

Metal Sample Showing Flourescence

flux

1. material used to remove oxide films from the surface of metals in preparation for soldering.
2. the radiant, or luminous, power

of a light beam; the time rate of the flow of radiant energy across a given surface.

■ flywheel effect

the continuous interchange of electric energy between the capacitor and inductor of a parallel resonant circuit wherein the energy level is diminished only by circuit resistance and radiated energy.

■ FM Frequency Modulation

where voltage levels change the frequency of a carrier wave.

■ folded resonator

construction in which the interior optical path is bent by mirrors mounted on corner blocks bolted into pre-aligned position, permitting compact packaging of a long laser cavity.

■ forward bias

a PN junction bias which allows current to flow through the junction. Forward bias decreases the resistance of the depletion layer.

■ FPGA

Field Programmable Gate Array; an integrated circuit fabrication technology in which pre-fabricated circuit modules are electrically configured by the user to meet specific design requirements on a chip-by-chip basis.

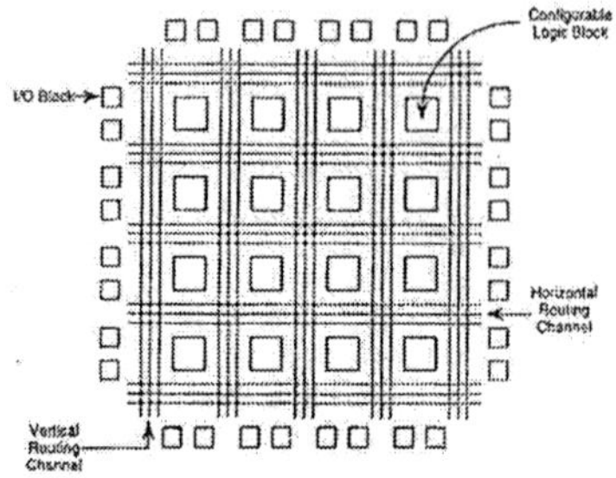

Field Programmable Gate Array

■ frame

a data structure that consists of one or more elements corresponding to an operation's defining attributes.

■ free runing multivibrator

a multivibrator that produces a continuous output waveform without any signal input. a square wave generator used to produce a clock signal.

■ frequency division

the use of circuitry to produce a reduction in the repetition rate of pulse waveforms or a decrease in the frequency of a-c signals.

■ frequency meter

meter used to measure frequency of periodic waves.

■ frequency modulation

1. a modulating system wherein the frequency of the carrier signal is shifted above and below

its normal frequency by the modulating signal.
2. modulation where the frequency of the sinewave carrier alters with the amplitude of the modulating signal.

■ **frequency multiplier**
a harmonic conversion circuit in which the frequency of the output signal is an exact multiple of the input frequency.

■ **frequency response curve**
1. a graph of amplitude over frequency indicating a circuit response to different frequencies.
2. indication of how well a circuit responds to different frequencies applied to it.

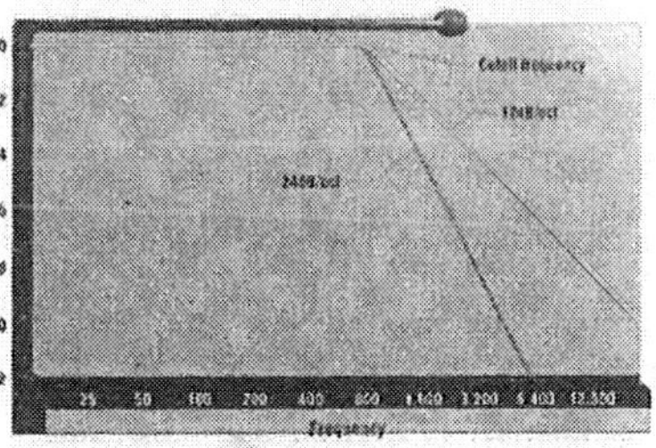

Frequency Response Curve of a Low Pass Filter

■ **frequency-division multiplex**
(FDM) transmission of two or more signals over a common path by using a different frequency band for each signal.

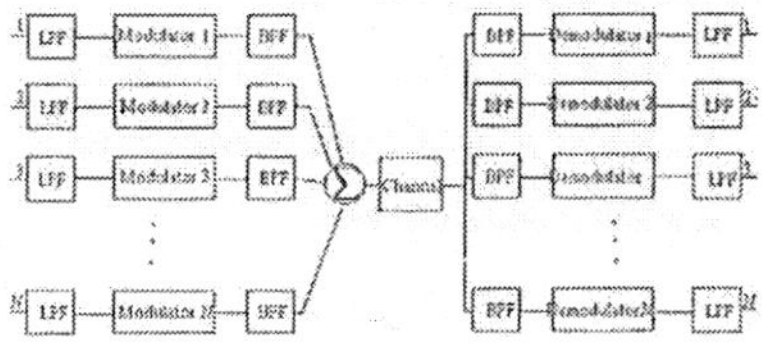

■ **frequency-domain analysis**
a method of representing a waveform by plotting its amplitude against frequency.

■ **front end**
in semiconductor manufacturing, front end refers to the fabrication process in which the integrated circuit is formed in and on the wafer.

■ **full duplex**
full-duplex communication means that data flow can be simultaneously transmitted and received across a communications channel.

■ **full scale deflection**
(FDS) deflection of a meter's pointer to the farthest position on the scale.

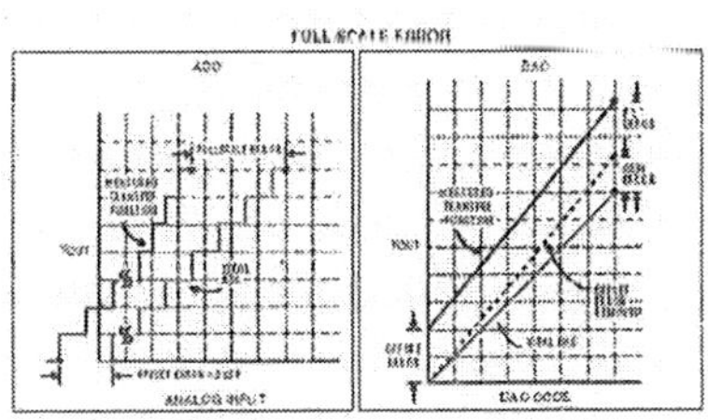

full wave rectifier

rectifier that makes use of the full AC wave in both the positive and negative half cycles.

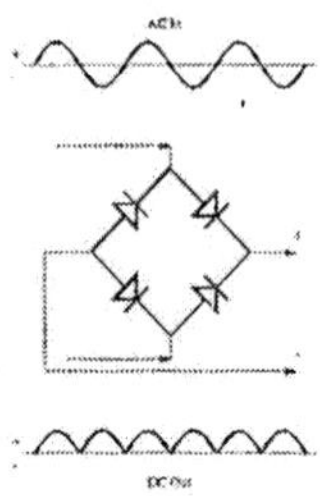

Full Wave Rectifier and it's Response to AC Voltage

full-power bandwidth (FPBW)

a large-signal (i.e. at or near full scale) analog input is applied to the analog-to-digital converter (ADC), and the input frequency is swept up to the point where the amplitude of the digitized conversion result has decreased by -3dB. This point is defined as the "full-power bandwidth frequency."

full-scale (FS) error

for the ideal transfer curve, the code edge transition that causes all ones in the digital code to occur 1. 5 LSB below full scale. The full-scale error is the difference between this code transition of the ideal transfer function and the actual measured value at this code transition. Unlike gain error, offset error is included in the FS error measurement.).

function call driver

a type of Keithley MetraByte board driver which provides a high-level alternative to register-level programming.

function generator

signal generator that can produce sine, square, triangle and sawtooth output waveforms.

fundamental frequency

lowest frequency in a complex waveform.

fuse

a protective device in the current path that melts or breaks when current exceeds a predetermined maximum value.

fusion splicer

an instrument that splices fibers by fusing or welding them, typically by electrical arc.

GaAs

Gallium Arsenide; a semiconductor crystal which is a compound of the elements gallium and arsenic. Its electronic properties offer the potential for higher speed electron conduction than in silicon, but at the expense of greater power consumption.

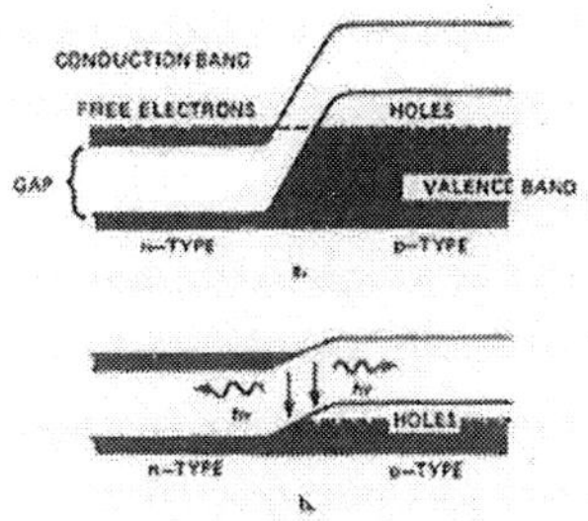

Energy-level Diagram of GaAs Semiconductor diode

■ gain

1. increase in voltage, current and/or power. Gain is expressed as a ratio of amplifier output value to the corresponding amplifier input value.
2. the ratio of the output level of a circuit to the input. This will be positive for an amplifier and negative for an attenuator.
3. a logarithmic measure of how well an antenna radiates or receives when compared to a reference antenna (a perfect point source dBi or a half wave dipole dBd), this can be highly directive with the intensity varying greatly over small angles.

■ gain bandwidth product

a device parameter that indicates the maximum possible product of gain and bandwidth. The gain bandwidth product of a device is equal to the *unity gain frequency (funity)* of the device.

■ gain error

with a full-scale analog input voltage applied to the analog-to-digital converter (ADC) (resulting in all ones in the digital code) or a digital code of all ones applied to the digital-to-analog converter (DAC), gain error is defined as the amount of deviation between the ideal transfer function and the measured transfer function (with the offset error removed). Gain error is usually expressed in LSB or a percent of full-scale range (%FSR). It can be calibrated out with hardware or in software.

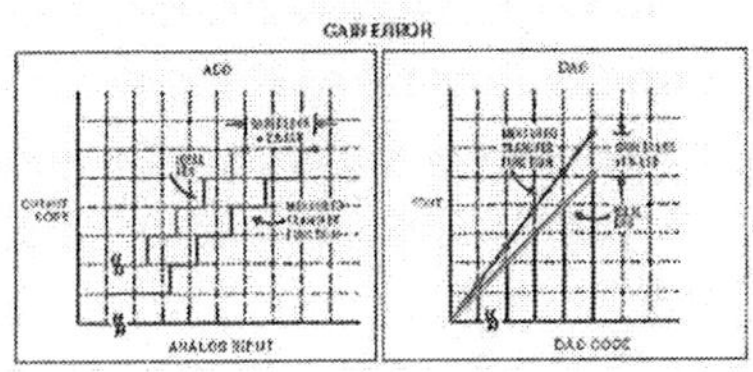

■ gas discharge laser

a laser containing a gaseous lasing medium in a glass tube in which a constant flow of gas replenishes the molecules depleted by the electricity or chemicals used for excitation. The discharged gas can be filtered and 90% recycled for economy.

■ gas jet assist

an assisting coaxial gas, such as oxygen, argon, or nitrogen, which may be used to achieve very high power levels for cutting certain metals.

■ **gas transport**
a laser design which generates very high beam power within a fairly small resonator structure. Long electrodes parallel the axis and gas is circulated across the resonator cavity.

■ **gate**
element of a digital circuit. Also one of the terminals of a Field Effect Transistor (FET).

■ **gate leakage current**
because the gate is reverse biased with respect to the channel, the current flowing to the gate terminal is very small; it is the reverse bias leakage current of a p-n junction.

■ **gate oxide**
a parameter related to reliability. Integrated Fowler-Nordheim current through insulator up to "breakdown."

■ **gate oxide**
the thin, high quality silicon dioxide layer in the MOSFET gate region.

■ **gate terminal**
along with the source and drain, one of the three terminals of an FET. This term also applies to the MIS capacitor, which has only two terminals the gate and substrate.

■ **gate signal**
a signal which in the active state enables an operation to occur; and when in the inactive state inhibits an operation from occurring.

■ **gated-beam detector**
an FM demodulator utilizing a single tube in a highly sensitive circuit.

■ **gating circuit**
a circuit which passes certain signals only when gating signals are applied.

■ **Gaussian**
the "normal curve," or normal distribution, an example of which is the symmetrical bell shape of the holes created by the uncorrected, unfocused laser beam in its optimum mode. a Gaussian laser beam has most of its energy in the centre.

■ **Geiger counter**
device used to detect nuclear particles.

■ **General Purpose Timer**
Usually an R-C oscillator providing a square wave output, with various control signals to perform Reset and gating (on/off)

functions. Control signals allow generation of single pulses (monostable), continuous pulses (astable), or a burst of a fixed number of pulses at the output.

■ giga
metric prefix for 1 billion (10^9).

■ Gilbert cell
an analog multiplier configuration useful for implementing mixers, up and down converters.

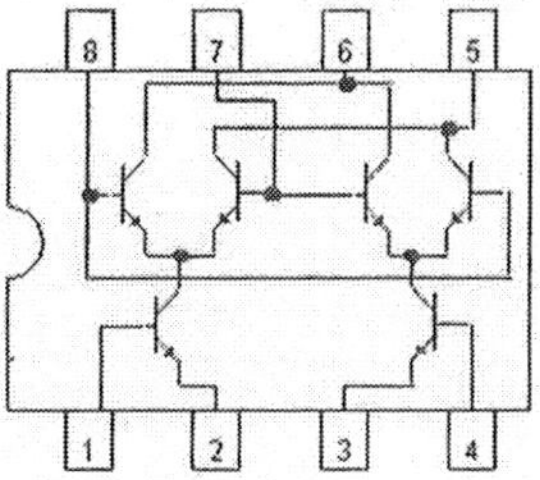

An Array of Gilbert Cell in an IC

■ GPIB
abbreviation for General Purpose Interface Bus. It is a standard for parallel interfaces.

■ graded index
a type of multimode fiber which used a graded profile of refractive index in the core material to correct for dispersion.

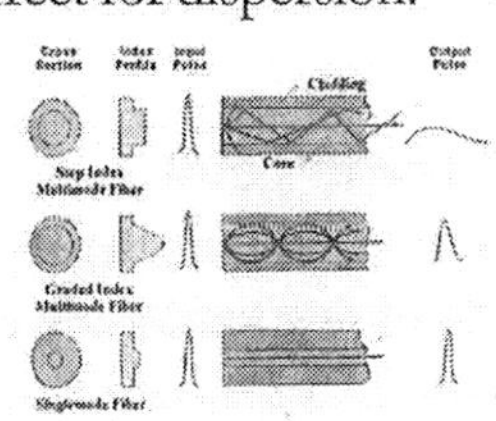

Comparision Between GIF and Other Fibers

■ gray code
unit-distance code system in which only one code signal changes with the transition from one measuring step to the next.

■ grid array
a group of elements (in a straight line or parabola for example) used to increase performance of a feed element (a Yagi antenna is one particular style of grid array).

■ grid modulation
the application of a modulating signal to vary the grid bias potential.

■ ground
an intentional or accidental conducting path between an electrical system or circuit and the earth or some conducting body acting in place of the earth. a ground is often used as the common wiring point or reference in a circuit.

■ ground plane
the earth or negative rail of a circuit. a large or significant mass that presents the effect of earth (ground) to a signal.

■ grounded base
a transistor amplifier circuit comparable to the grounded-grid (signal ground) tube circuit.

■ **grounded collector**

a transistor circuit comparable to the cathode-follower tube circuit. (also called emitter follower.).

■ **grounded c-v measurement**

a capacitance vs. voltage measurement in which one terminal of the capacitor is grounded. typically, this means the capacitor gate is stimulated with the test signal and the response is also measured from the gate.

■ **grounded emitter**

a transistor circuit comparable to the conventional grounded cathode tube amplifier.

■ **grounded grid**

a tube circuit which has the control grid placed at signal ground.

■ **GSM**

Global System for Mobile Communications; GSM has become the world's most widely used mobile system, operating on the 900 MHz and 1800 MHz frequencies in Europe, Asia and Australia, and the 1900 MHz frequency in North America and Latin America.

■ **guarding**

a technique that reduces leakage errors and decreases response time. consists of a guard conductor driven by a low-impedance source surrounding the lead of a high-impedance signal. the guard voltage is kept at or near the potential of the signal.

■ **gunn diode**

a semiconductor diode that utilises the gunn effect to produce microwave frequency oscillation or to amplify a microwave frequency signal.

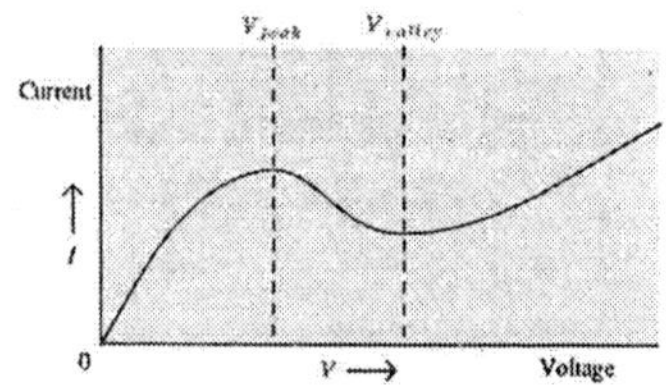

I/V curve of a Gunn diode.

■ **H filter**

a filter circuit containing five components so wired to resemble the letter h.

■ **half duplex**

half-duplex communication means that data flow can be alternately transmitted and received across a communications channel.

■ **half power point**

1. a frequency at which the power is 50% of maximum. this corresponds to 70.7% of maximum current or voltage.

2. the points on a frequency-response curve at which the output power is one- half of the maximum power out.

■ **half wave rectifier**
a diode rectifier that converts ac to pulsating dc by eliminating either the negative or the positive alternation of each input ac cycle.

■ **Hall effect**
the measurement of the transverse voltage across a conductor when placed in a magnetic field. by this measurement, it is possible to determine the type, concentration, and mobility of carriers in silicon.

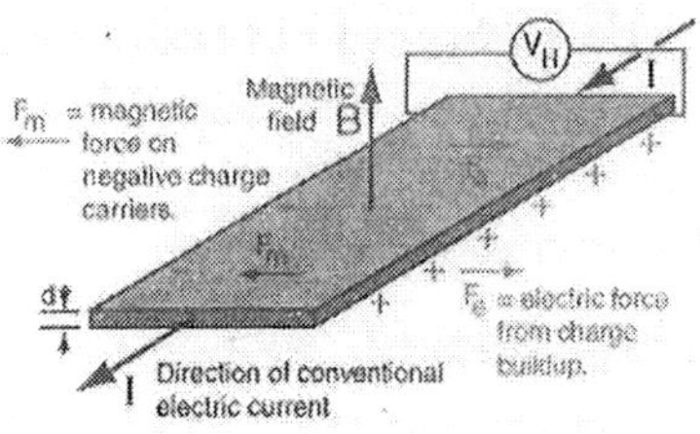

Hall Effect Shown by a Mettalic Strip

■ **Hanning Window**
FFT window function that normally provides better frequency resolution than the flat top window, but with reduced amplitude accuracy.

■ **hardware**
the actual integrated circuit chips and interconnections in a system as opposed to the computer programs and data (software) which may be stored in memory (which is itself hardware).

■ **harmonic**
sine wave that is smaller in amplitude and some multiple of a fundamental frequency. example: 880 Hz. is the second harmonic of 440 Hz., 880 Hz. is the third harmonic of 220 Hz.

■ **Hartley oscillator**
a signal generator circuit in which the inductor of the resonant circuit is tapped by the cathode of the oscillator tube or the emitter of the transistor.

■ **Hay Bridge**
a bridge circuit in which two opposite arms are resistors, the third arm c and b in series, and the fourth arm l and r in series.

■ **HDL**
Hardware Description Language; a human and computer-readable language used to describe the functions and/or behavior of an integrated circuit or system which can be used to carry out computer simulation or automatically synthesise the actual circuits necessary to implement the specified functions.
heat sink.

a substance or device used to dissipate or absorb unwanted heat, as from a manufacturing process (or, with lasers, from reflected rays).

■ heater

a vacuum-tube electrode which heats the cathode. in such an indirectly heated tube, the filament (heater) does not serve as the electron source.

■ heavy spot

the angular location of the imbalance vector at a specific lateral location on a shaft. the heavy spot typically does not change with rotational speed.

■ helium-neon laser,

("HeNe"), Laser in which the active medium is a mixture of helium and neon, which is in the visible range. Used widely in industry for alignment, recording, printing, and measuring, it is also valuable as a pointer or aligner of invisible CO_2 laser light.

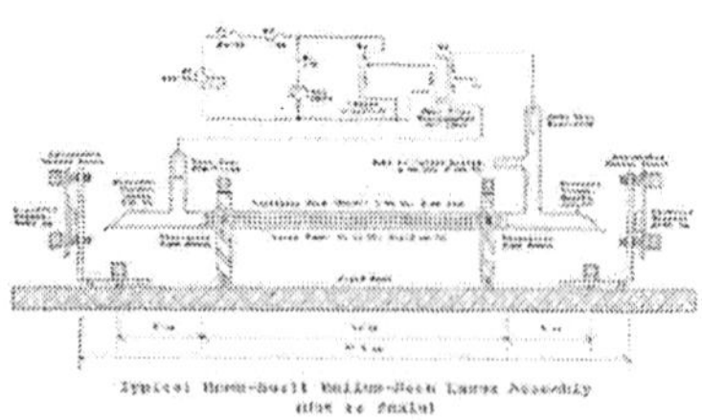

Helium ION Structure

■ Henry

the electromagnetic unit of inductance. 1 Henry occurs when a back emf of 1 volt is produced when current is changing at the rate of 1 amp per second.

■ Hertz ,Hz

the unit of frequency in cycles per second.

■ heterodyne

the electronic mixing of two signals of different frequency to produce a third signal having a frequency equal to the difference in frequency between the original signals.

■ high fidelity

(Hi Fi)Sound reproduction equipment that reproduces sound as near to the original sound as possible.

■ high fidelity

an amplifying system which faithfully reproduces virtually a full audio-frequency spectrum with negligible distortion (approximately 20 cps to 20,000 cps).

■ high frequency c-v

a capacitance vs. voltage measurement obtained using such a high frequency test signal that neither minority carriers in the inversion layer nor interface traps can follow the ac gate voltage. thus, they do not contribute to the high frequency capacitance.

■ high spot

the angular location on the shaft directly under the vibration transducer at the point of closest proximity. the high spot can move with changes in shaft dynamics (e.g., from changes in speed).

■ high-pass filter

1. a tuned circuit designed to pass all frequencies above a designated cut-off frequency. frequencies below the cut-off frequency are rejected or attenuated

2. a filter that allows only signals above a certain frequency through. (all frequencies below that certain frequency are attenuated.) the characteristic frequency is often referred to as the corner frequency.

■ hole

a gap left in the covalent bond when a valence electron gains sufficient energy to jump to the conduction band.

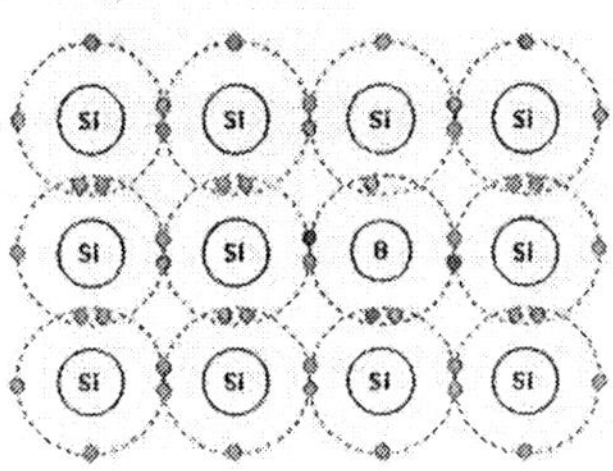

Creation of Hole

■ hole flow

current carriers (in transistors) which have a flow in a direction opposite to electron movement.

■ hologram

three-dimensional picture created with a laser.

■ holography

the science dealing with three-dimensional optical recording.

■ hot carrier degradation

a parameter related to reliability. accelerated test to predict lifetime. typical indicator is change in IDSAT.

■ hot chuck

a platform or chuck, mounted on a prober onto which a wafer is placed for testing, which can be heated (typically to 150-300°c) for testing at elevated temperatures.

■ hot junction

the junction of two dissimilar metals in a thermocouple circuit that is used to measure an unknown temperature. also known as measurement junction.

■ h-parameters

(hybrid parameters) Transistor specifications that describe the component operating limits under specific circumstances.

hybrid circuit

circuit that combines two technologies (passive and active or discrete and integrated components) onto one microelectronic circuit. passive components are usual made by thin film techniques, while active components are made with semiconductor techniques.

hydrophone

a hydrophone converts acoustic energy into electrical energy and is used in underwater, passive systems for listening only. hydrophones are usually used below their resonance frequency over a much wider frequency band where they provide uniform output levels.

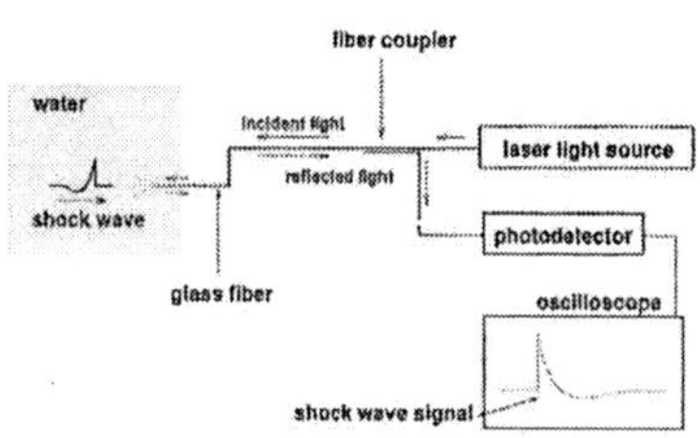

Working of Hydrophone

hydrophone directivity

the beam width of a hydrophone determines its directivity. a narrow beam will give it greater directivity, i.e. allow determination as to the direction a sound wave is coming from.

hysteresis

amount that the magnetization of a material lags the magnetizing force due to molecular friction. in Schmitt trigger circuits, the difference between the upper and lower trigger points.

IC

integrated Circuit; the interconnection of multiple electronic circuit components fabricated on a single semiconductor crystal. Often used interchangeably with chip.

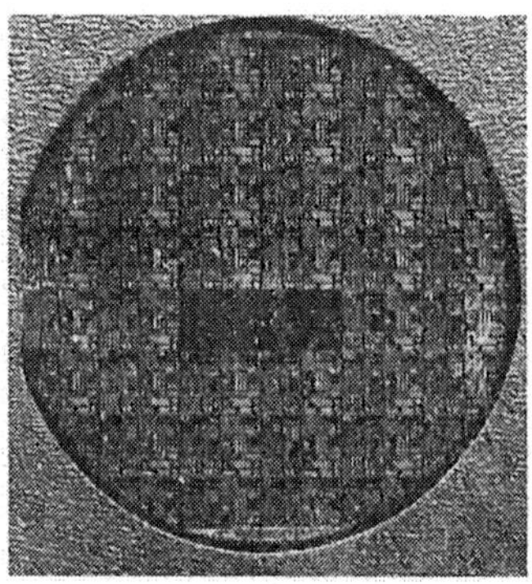

IC Layout on a Wafer

IC voltage regulator

three terminal device used to hold the output voltage of a power supply constant over a wide range of load variations.

IEEE

abbreviation for institute of electrical and electronics engineers.

IGFET
insulated gate field effect transistor. Another name for a "MOSFET."

ignitron
a mercury-pool type of rectifier tube employing mercury-vapor ionization.

image
the optical reproduction of an object, produced by a lens or mirror. a typical positive lens converges rays to form a "real" image which can be photographed. a negative lens spreads rays to form a "virtual" image which can't be projected.

impedance
similar to resistance but applies to ac circuits. the opposition a component in a circuit offers to the flow of alternating current at a given frequency. if the frequency changes, the impedance changes too. impedance is measured in ohms and has the symbol "z" with unit ohm.

impedance coupling
coupling of two signal amplifier circuits through the use of an impedance such as a inductor.

impedance matching
matching the output impedance of a source to the input impedance of a load to attain maximum power transfer.

impulse response
the response of a system to an impulse as input signal. the output then produces the impulse response that is the time domain equivalent to the frequency response function.

Impurity
a material that is present in minute quantities in an ideally pure material. in the context of C-V testing, this term is synonymous with dopant.

in phase
when two or more waves of the same frequency have their positive and negative peaks occurring at the same time.

incandescence
state of a material when heated to the point where it emits light. (red hot or white hot).

incident light
a ray of light that falls on the surface of a lens — or any other object. the "angle of incidence" is the angle made by the ray with a perpendicular to the surface.

■ **Incremental measuring system:**
measuring method by which the measuring value is derived by the summation (counting) of increments (measuring steps).

■ **Index matching fluid**
a liquid used of refractive index similar to glass used to match the materials at the ends of two fibers to reduce loss and back reflection.

■ **Index of refraction**
a measure of the speed of light in a material.

■ **index profile**
the refractive index of a fiber as a function of cross section.

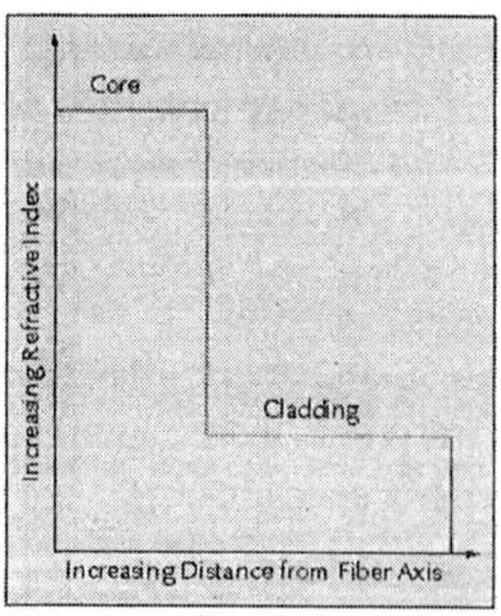

refractive index profile, multimode step-index fiber

■ **indirectly-heated tube**
a tube in which the filament heats the cathode to produce electron emission.

■ **induced voltage**
voltage generated in a conductor when subjected to a moving magnetic field.

■ **Inductance**
a changing current in a coil produces a changing magnetic flux. the changing magnetic flux results in an induced current flowing in the coil. the unit is a henry. One henry is the inductance of a circuit that produces a counter-force of 1 volt when the current flowing is changing at the rate of 1 amp per second.

■ **Induction**
the process of inducing by magnetic lines of force a potential or magnetization in another component.

■ **inductive circuit**
circuit having greater inductive reactance than capacitive reactance.

■ **inductive reactance**
opposition to the flow of ac current produced by an inductor. measured in ohms and varies in direct proportion to frequency.

■ **inductive reactance**
the opposition which an inductor offers to ac for a given signal frequency.

inductor
length of conductor used to introduce inductance into a circuit. The conductor is usually wound into a coil to concentrate the magnetic lines of force and maximise the inductance. While any conductor has inductance, in common usage the term inductor usually refers to a coil.

influence coefficients
mathematical coefficients that describe the influence of system loading on system deflection.

infrared
1. electromagnetic heat radiation whose frequencies are above the microwave frequency band and below red in the visible band.
2. infra-red is a type of light wave. We cannot see it because it is just outside the range of light which our eyes can detect. You can feel it, though - infra-red light warms things and is the way an electric fire heats a room. The infra-red equipment used in the projects we have designed is not that powerful and you will not feel any heat from it.

inhalation
taking a material into the body by breathing it in.

inhibit
to stop an action or block data from passing.

inhibitor circuit
a circuit in which the application of one signal prevents (inhibits) the transfer through the circuit of another signal.

input bias current
the amount of current that flows into or out of an analog input pin. bias current will add to the error of a measurement as it interacts with the input signal source impedance.

input bias current
the current that flows at the input due to internal circuitry and bias voltage. also, the current that must be supplied to the high input measuring terminal (with zero input signal and offset voltage) to reduce the output indication to zero.

input impedance
1. opposition to the flow of signal current at the input of a circuit or load.
2. the amount of load an input places on a signal source that drives it. high input impedance is very often desired and implies that little change in the signal is expected when the circuit is connected.

input isolation
on a switching card, the isolation from signal high to low (or guard) for a two-pole circuit. Specified as resistance and capacitance.

input offset current
the difference in bias current when measured between two differential input pins such as the (+) and (-) inputs of an amplifier.

input voltage range
the maximum and minimum signal that can be applied to an analog input for rated performance.

input/output (I/O)
the process of transferring data to and from a computer-controlled system using its communication channels, operator interface devices, data acquisition devices, or control interfaces.

insertion loss
the attenuation of signals due to routing them through a switching card. specified as a decibel value over a frequency range.

instantaneous value
the amplitude of a waveform at a given instant of its duration.

instrument amplifier
a circuit that amplifies a signal in a precise way while adding little error. it is generally implied that the amplifier has differential inputs and high input impedance.

insulated
when a non conducting material is used to isolate conducting materials from one another.

insulating material
material that will prevent the flow of current due to its chemical composition.

insulation resistance
resistance of insulating material. the greater the insulation resistance, the better the insulation.

integral coupling
innovative angle encoder design with built-in coupling, located preferably on the stator-side.

integral nonlinearity (INL) error
the amount of deviation of the measured transfer function of an analog-to-digital converter (ADC) or a digital-to-analog converter (DAC) from the ideal transfer function (defined as a straight line drawn from zero to full scale). (sometimes a "best-fit" straight line is used where the ideal transfer function is represented

by a straight line drawn between the end points of the actual transfer function. this method is referred to as "end-point linearity.") the INL error is also defined as the sum of the DNL errors starting from code 000 to the code where the INL measurement is desired.

integrated

when two or more components are combined into a circuit and then incorporated into a single package.

integrated circuit

an integrated circuit (IC) is a complete, very compact circuit made up from tiny resistors, capacitors, diodes and transistors. each IC can contain hundreds of these components, and the circuits themselves are very complicated. however, we do not need to know the exact details, just what the IC needs as inputs and what we can expect as outputs. if you look in an electronics catalogue, you will usually only be told what voltage the IC was designed for and how each pin of the IC must be connected if it is to work properly. one IC commonly used in school is the 555 timer.

integrating conversion

an analog to digital conversion process where the output results in a digital representation of the integral of the input signal over a specified time interval.

Integration

a process producing a result that, when differentiated, yields the original quantity. integration of acceleration, for example, yields velocity. integration is performed in a dsa by dividing the frequency lines by jw, where w is frequency multiplied by 2p.(integration is also used to convert velocity to displacement.)journal-specific portions of the shaft surface from which rotor applied loads are transmitted to bearing supports.

integrator

a device that approximates and whose output is proportional to an integral of the input signal. a low pass filter.

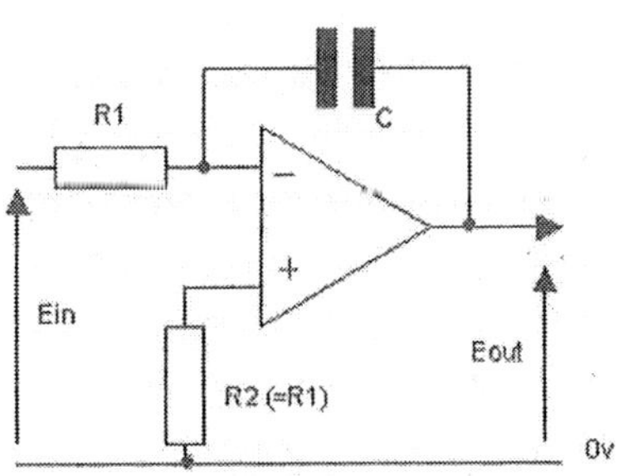

An Integrator Circuit

intensity
the magnitude of radiant energy (light) per unit, such as time or reflecting surface.

Interconnect
the metal and silicon patterns on an integrated circuit which serve to electrically interconnect components (e. g. transistors) and to supply power and distribute electrical data throughout the circuit.

interelectrode capacity
the internal capacitances existing between the elements of a vacuum tube.

Interface
usually refers to the hardware that provides communication between various items of equipment.

interface trap density
the probability per unit area that an interface trap level is present at a particular energy. the units are cm^{-2} ev^{-1}.

interface trapped charge
charge located on traps (defects, impurities) at the semiconductor-insulator interface. the trapping of charge otherwise involved on the operation of an mis device lowers its efficiency, degrading such characteristics as gain, channel leakage, and noise.

Interference
a disturbance to the signal in any communications system.

interferential measuring system:
photoelectric measuring system with a phase grating scale where scanning
signals are produced via the interference of diffracted beams.

Interlevel oxide
interlevel oxide is a layer of oxide that separates metal layers.

interlevel shorts. (parametric testing)
a yield parameter. estimated number of interlevel shorts per unit overlap area.

intermediate frequency
the signal obtained by heterodyning two signals of different frequencies. in superheterodyne receivers, the intermediate frequency is the difference between the incoming signal frequency and the local oscillator frequency.

intermediate frequency amplifier
in a superheterodyne radio it amplifies a fixed frequency lower than the received radio frequency and

higher than the audio frequency.

intermittent

a fault occurring at random intervals of time. intermittent problems are often difficult to locate because of the random nature. they often don't occur when the technician is present.

intermodulation distortion (IMD)

an analog-to-digital converter (ADC) test where two signals (f1 and f2) of equal amplitude that are also very close in frequency are applied to the ADC input. a frequency spectrum plot of the data reveals the amount of distortion due to the ADC's digitising of the two signals. IMD is typically specified in decibels. where the distortion terms are given by

2nd-order terms

f1 + f2, f1-f2

3rd-order terms

2f1 + f2, 2f1-f2, f1 + 2f2, f1-2f2

internal resistance

every source has some resistance in series with the output current. when current is drawn from the source some power is lost due to the voltage drop across the internal resistance. usually called output impedance or output resistance.

internal trigger

a software-generated event that starts an operation. see also.

Interpolation

method for producing measuring increments which are smaller than a

fourth of the scanning signal period.

interrupt level

a specific priority that ensures that high priority interrupts get serviced before low priority interrupts.

interrupt service routine (ISR)

a software program that handles interrupts.

interrupt-mode operation

mode in which a data acquisition board acquires or generates samples using an interrupt service routine (ISR).

intrinsic material

a semiconductor material with electrical properties essentially characteristic of ideal pure crystal. essentially silicon or germanium crystal with no measurable impurities.

■ **intrinsic semiconductor**
a pure semiconductor that has not been doped. since it is not doped, it is neither n-type nor p-type and electrons and holes carry charge in equal concentrations.

■ **intrinsic stand-off ratio**
a unijunction transistor (UJT) rating used to determine the firing potential of the device.

■ **inverse feedback**
a circuit which feeds back a portion of an amplified signal to a previous stage in an out-of-phase condition with respect to the signal at that stage to improve signal-to-noise ratio and frequency response. also called negative feedback.

■ **inversion region**
the region of a c-v curve where high frequency capacitance is low and quasistatic is high. in this region, majority carriers are repelled from the semiconductor-insulator interface and minority carriers are attracted to it, building up and inverting the local conductivity type.

■ **inverter**
a circuit in both analog and digital systems that provides an output that is inverse to the input. also a circuit that converts dc to ac.

■ **inverting amplifier**
an amplifier that has a 180° phase shift from input to output.

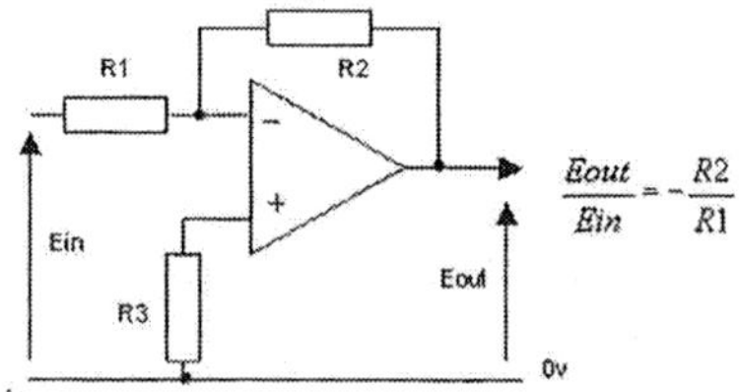

■ **inverting input**
in an operational amplifier (op amp) the input that is marked with a minus sign. a signal applied at the inverting input will be given 180° phase shift between input and output.

■ **ion**
an atom with fewer electrons in orbit than the number of protons in the nucleus is a positive ion. an atom with a greater number of electrons in orbit than the number of protons in the nucleus is a negative ion.

■ **ion implantation**
introduction of dopant ions into a semiconductor material by accelerating them through a high voltage electric field and bombarding the semiconductor material. this doping process results

in a layer of dopants buried in the semiconductor.

ion laser

a type of laser employing a very high discharge current, passing down a small bore to ionise a noble gas such as argon or krypton. the ionization process creates a population inversion for lasing to occur. a research laser useful for some industrial applications.

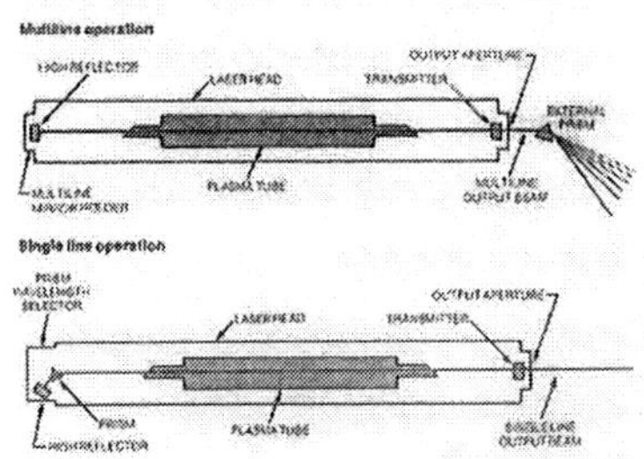

Single and Multiline Operation of ION Laser

ion

this is what an atom becomes when an electron is separated from the atom, leaving it with a net positive charge or, if an electron is added, leaving it with a net negative charge.

ionisation

in gas tubes, the bombardment of gas atoms by the electron stream which removes electrons from the gas atoms to form ions. the process increases conduction and reduces internal resistance.

ISDN

Integrated Services Digital Network.

isolated outputs

output signals where a common reference is not connected to either input terminal.

isolation leakage. (parametric testing)

a CMOS device parameter. isolated junction to junction current at maximum reverse voltage.

a prefix used to indicate an imaginary number. (Operator j).

j operator

in rectangular notation, a symbol representing a 90° rotation and equal to *i (square root of -1)*. the j values represent reactance.

jack

socket or connector into which a plug may be inserted.

jacket

the protective outer coating of the cable.

JFET

abbreviation for "Junction field effect transistor".

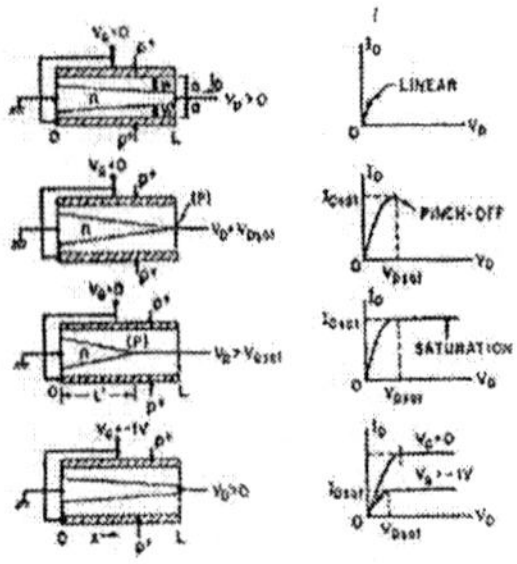

JFET Working Diagram

■ **joule**

the unit of work and energy.

■ **jumper cable**

a short single fiber cable with connectors on both ends used for interconnecting other cables or testing.

■ **jumper cable**

a short single fiber cable with connectors on both ends used for interconnecting other cables or testing.

■ **junction**

contact or connection between two or more wires or cables. the area where the *p*-type material and *n*-type material meet in a semiconductor.

■ **junction area capacitances. (parametric testing)**

a CMOS or bipolar device parameter. capacitance per unit area, as function of voltage.

■ **junction area leakages. (parametric testing)**

a CMOS or bipolar device parameter or a yield parameter. junction leakage per unit junction area.

■ **junction diode**

a semiconductor diode in which the rectifying characteristics occur at a junction between the *n*-type and *p*-type semiconductor materials.

■ **junction FET (JFET)**

a type of FET in which the gate forms a Schottky contact with the semiconductor material. c-v is not generally measured on this type of device.

■ **junction perimeter capacitances. (parametric testing)**

a CMOS or bipolar device parameter. isolation-sided or emitter/collector-sided leakage per unit length.

■ **junction perimeter leakages. (parametric testing)**

a CMOS or bipolar device parameter. leakage per unit isolation-sided and gate-sided perimeter.

■ **junction perimeter leakages (parametric testing)**

a yield parameter. isolation-sided

or emitter/collector-sided leakage per unit length.

■ **Kelvin contacts**
a means for testing or making measurements in electronic devices and circuits, particularly when low values are being measured. two sets of leads are used at each test point, similar with respect to thickness, material and length; one set carries the test signal and the other connects with the measuring instrument. the effect of resistance in the leads is thus eliminated.

■ **keyholing**
in welding, the deep-penetration holes, which fill quickly with molten metal, that can be made in a few milliseconds by laser.

■ **keyphasor**
a signal used in rotating machinery measurements, generated by a transducer observing a once-.

■ **kHz (kilohertz)**
unit of frequency, equal to one thousand hertz or cycles per second.

■ **kilo**
metric prefix for 1000. (10^3).

■ **kilobyte (k)**
1024 bytes.

■ **kilocycles (kc)**
one thousand cycles.

■ **kilovolt-ampere**
1000 volts at 1 ampere.

■ **kilowatt - hour kwh**
a unit of energy when one kilowatt of power is expended for one hour. example a radiator bar is usually rated at 1,000 watts and this switched on for one hour consumes one kilowatt-hour of electricity.

■ **kilowatt-hour meter**
a meter used by electric utility companies to measure the amount of electric power used by a customer.

■ **kinetic energy**
energy associated with motion.

■ **Kirchhoff's current law**
the basic law which states that the sum of currents flowing into any junction of an electric circuit is equal to the sum of currents flowing out of the junction.

■ **Kirchhoff's voltage law**
the basic law which states that the sum of voltage sources around any closed circuit is equal to the sum of the individual voltage drops across the resistances of the circuit.

knee voltage

the voltage at which a curve joins two relatively straight portions of a characteristic curve. for a pn junction diode, the point in the forward operating region of the characteristic curve where conduction starts to increase rapidly. for a zener diode, the term is often used in reference to the zener voltage rating.

L Pad

an attenuator having components arranged to resemble the letter l.

ladder filter

a type of low pass filter. a ladder filter is among the most robust types (i.e., insensitive to parasitic effects and tolerances), but it is also most difficult to design.

lag

difference in time between two waveforms of the same frequency expressed in degrees. example: one waveform lags another waveform by a certain number of degrees.

laminated core

core made up of sheets of magnetic material insulated from one another by an oxide or varnish.

lamp

device that produces light.

LAN

Local Area Network; a small data network covering a limited area, such as within a building or group of buildings.

LASER

an acronym of light amplification by stimulated emission of radiation. a laser is a cavity that has mirrors at the ends and is filled with lasable material such as crystal, glass, liquid, gas, or dye. these materials must have atoms, ions, or molecules capable of being excited to a metastable state by light, electric discharge, or other stimulus. the transition from this metastable state back to the normal ground state is accompanied by the emission of photons which form a coherent beam.

LASER accessories

the hardware and options available for lasers, such as secondary gases, brewster windows, q-switches, and electronic shutters.

laser diode, ILD

a semiconductor device that emits high powered, coherent light when stimulated by an electrical current. used in transmitters for single mode fiber links.

laser hardening

laser-beam traversal of metal to harden quenching process producing the maximum hardness for most metals.

laser oscillation

the buildup of the coherent wave between laser cavity end mirrors. in cw mode, the wave bounding back and forth between mirrors

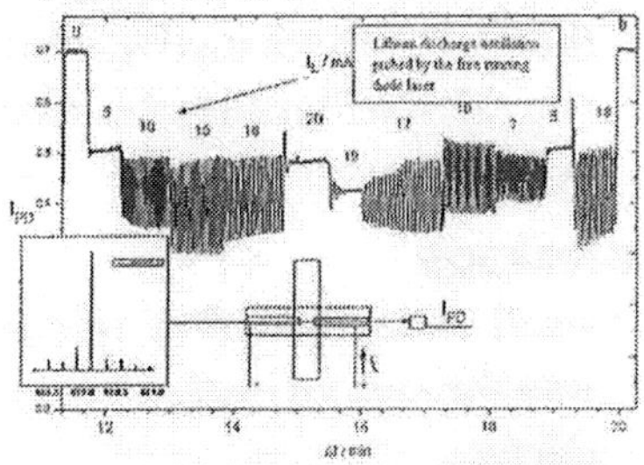

transmits a fraction of its energy on each trip; in pulsed operation, emission happens instantaneously.

laser rod

a solid-state, rod-shaped lasing medium in which ion excitation is caused by a source of intense light, such as a flash lamp. various materials are used for the rod, the earliest of which was synthetic ruby crystal.

latch-up

a parameter related to reliability. maximum current forced through any terminal before latch-up occurs.

lateral location

the definition of various points along the shaft axis of rotation.

launch cable

a known good fiber optic jumper cable attached to a source and calibrated for output power used used as a reference cable for loss testing. this cable must be made of fiber and connectors of a matching type to the cables to be tested.

layout

the physical location of the components and interconnections which make up an integrated circuit. the layout data include the geometrical patterns on each of the masks required during processing.

L-C oscillator

oscillator in which the frequency determining components are an inductor and capacitor.

LC50

the concentration of a material in air that causes death in 50% of a group of test animals. The material is inhaled over a set period of time, usually 4 hours. LC stands for lethal concentration.

LCD

Liquid Crystal Display - a reflective display that requires very low power for operation.

■ **LCZ meter**

inductance (l), capacitance (c), impedance (z) meter. a general purpose instrument for measuring component l, c. And z. Sometimes called lcr meter. This instrument may be applied to c-v testing, but typically lacks features optimised for c-v.

■ **LD50**

the weight of material that causes the death in 50% of a group of test animals. It is usually expressed in weight of material per weight of test animal. LD stands for lethal dose.

■ **LDR**

Light-Dependent Resistor. The resistance of the unit varies with changing intensity of incident light.

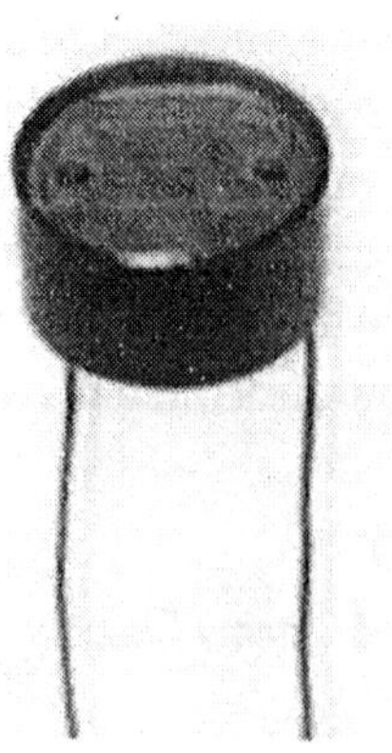

■ **lead**

the angle by which one alternating signal leads another in time. Opposite of lag. Also a wire that connects two points in a circuit.

■ **leading edge spike**

the initial pulse in a series of pulsed laser emissions, often useful in starting a reaction at the target surface. The trailing edge of the laser power is used to maintain the reaction after the initial burst of energy.

■ **leakage**

small undesirable flow of current through an insulator or dielectric.

■ **leakage current.**

error current that can degrade sensitive measurements. Leakage current is any unwanted current that flows when test voltage is applied. The ideal leakage current is zero. Leakage currents can originate in instruments, cables, or the device being tested. Even high resistance paths between low current conductors and nearby voltage sources can generate significant leakage currents.

■ **leakage resistance.**

the resistance calculated from the dependence of leakage current on applied voltage.

leakage resistance = V / leakage current.

Least Significant Bit (LSB)

In the binary number scheme, the bit position in a group of bits given the smallest weighted value. The LSB is typically the furthest right bit in a grouping of data bits. The LSB weight is given by the full-scale range of the converter divided by 2N, where N is the converter's resolution.

LED

Light Emitting Diode. a diode that emits light when current is passed through it. It has two leads cathode (k) and anode.

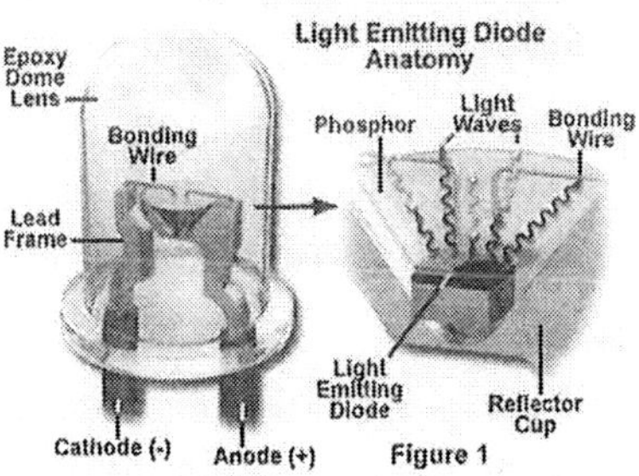

Internal View of a Typical LED

left-hand rule

the procedure for finding electron-flow direction in a wire by grasping it so the fingers point in the direction of the magnetic lines of force. The thumb now points to electron-flow direction. For a single-layer inductor, the thumb indicates the north-pole end of the coil when the fingers point in the direction of electron flow. For conventional current flow, the right-hand rule is used.

level control

a variable control for adjusting signal amplitude.

level detector

an op-amp circuit that compares two inputs and provides a dc output indicating the polarity relationship between the inputs.

level sensor

same as a proximity sensor except with the surface of a fluid or bulk solid as the target.

lie detector

piece of electronic equipment also called a polygraph used to determine whether a person is telling the truth by looking for dramatic changes in blood pressure, body temperature, breathing rate, heart rate and skin moisture in response to questions.

lifetime

the time from the creation of an electron hole pair until recombination occurs.

light

the range of electromagnetic radiation frequencies detected by the eye, or the wavelength range from about 400 to 750 nanometers. It is sometimes extended to include photovoltaic effects and radiation beyond visible limits.

■ **light dependant resistor, LDR**

this is a special type of RESISTOR and is sometimes called a PHOTOCONDUCTOR. LDRs are made so that their resistance decreases as the level of light falling on them increases. In the dark they have a high resistance in the light they have a low resistance. It is an input device.

■ **light emitting diode ,LED**

these DIODES emit light when current flows from ANODE to CATHODE. They normally emit red light, but other colour are available, including INFRARED LEDs. They conduct current in one direction only, but can be damaged by reverse voltages. They have to be connected the right way around in the circuit to work The smaller leg, or the leg adjacent to the flat edge on the top of the LED must be connected to negative or ground. LEDs are output devices.

■ **light regulation**

a form of power regulation in which output power is maintained at a constant level by controlling discharge current.

■ **limiter**

circuit or device that prevents some portion of its input from reaching the output. a clipper.

■ **line regulation**

the ability of a voltage regulator to maintain a constant voltage when the regulator input voltage varies.

■ **linear**

relationship between input and output in which the output varies in direct proportion to the input.

■ **linear scale**

a scale in which the divisions are uniformly spaced.

■ **linearity**

the response characteristics of a linear system remain constant with input level and/or excitation signal type. that is, if the response to input a is k. a, and the response to input b is k. b, then the response of a linear system to input(a + b) will be (k. a + k. b), independent of the function k. an example of a nonlinear system is one whose responses limited by mechanical stop, such as occurs when a bearing mount is loose.

■ **local buffer**
temporary memory location within an application program's memory area. It is always available to the application program.

■ **logic**
science of dealing with the principle and applications of gates, relays and switches.

■ **logic gate**
a circuit module which performs a common logical function such as inversion or logical AND and OR operations.

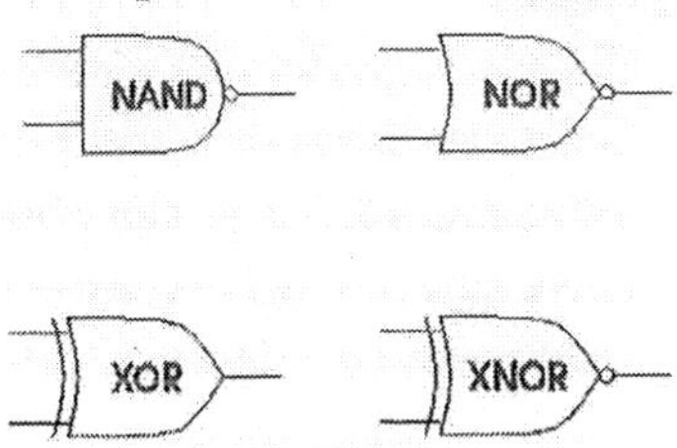

Logic Gates

■ **long-term accuracy**
the limit that errors will not exceed during a 90-day or longer time period. It is expressed as a percentage of reading (or sourced value) plus a number of counts over a specified temperature range.

■ **low-pass filter**
a filter that allows only signals below a certain frequency through. (all frequencies above that frequency are attenuated.) the characteristic frequency is often referred to as the corner frequency.

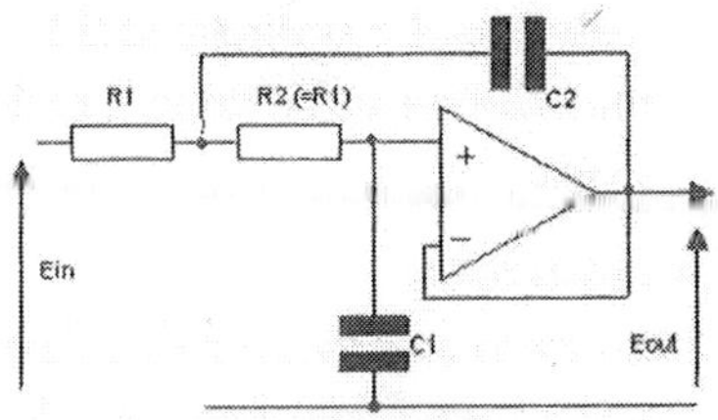

$R1 = R2 = R \quad C2 = \frac{\sqrt{2}}{2\pi f_c R} \quad C1 = \frac{C2}{2}$

An OP-AMP LPF

■ **magnetic amplifier**
a device using a saturable reactor capable of controlling the amount of power applied to a load by very small signal power applications to the control section.

■ **magnetic circuit breaker**
circuit breaker that is tripped or activated by use of an electromagnet.

■ **magnetic coil**
spiral of a conductor which is called an electromagnet.

■ **magnetic core**
material that exists in the centre of the magnetic coil to either physically support the windings (non-magnetic material) or to concentrate the magnetic flux (magnetic material).

magnetisation curve

shows the amount of magnetisingforce required to saturate a ferromagnetic material. it is normally shown as a graph with b as the ordinate and h as the abscissa,and is known as the B-H curve.

main lobe

the main acoustic beam in a directional transducer. there are other, smaller lobes called side lobes that are located around the main lobe.

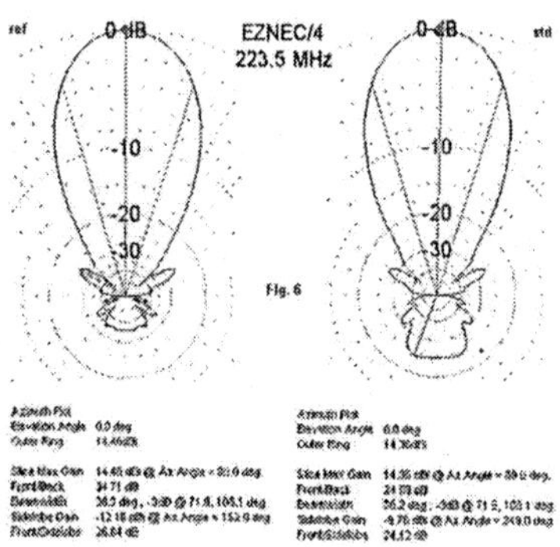

Pattern of Main Lobe

major carry transition

the mid scale point where the MSB changes from low to high and all other bits change from high to low, or where the MSB changes from high to low and all other bits change from low to high. this point is often where the worst switching noise occurs.

majority carrier correction

a method for measuring doping profiles that corrects for the onset of accumulation at shallow depths. Also known as "the Ziegler method" after its creator.

majority carrier

the type of free charge carrier that predominates in an extrinsic semiconductor material. For n-type, these are electrons (N means negative). For p-type, these are holes (P means positive).

make-before-break

connecting a new circuit before disconnecting the present circuit.

mask

the photographic film used to isolate portions of a wafer during a processing step. This photomask may be used to control the patterning of a layer of photo resist or other material on the surface of the wafer which acts as the actual process mask.

master

a mainframe that has control of other mainframes (slaves) through an external connection. a slave unit adds capacity or functions to the master. The master/slave combination has one IEEE-488 bus address.

■ **matched**

the condition existing when the load (antenna) is a pure resistance (no reactance) whose value equals the characteristic impedance of the line. The characteristic impedance of the line must also match the output impedance of the transmitter (typically 50 ?). Maximum transfer of power from radio to antenna requires a matched condition.

■ **matrix card**

a type of card with a switching configuration that has columns and rows of relay cross points. (Also called a coordinate switch.) With a matrix card, since any one point can be connected to any other, you can simultaneously have one input with multiple outputs, multiple inputs with one output, or multiple inputs with multiple outputs.

■ **maximum allowable input**

the maximum dc plus peak ac value (voltage or current) that can be applied between the high and low input measuring terminals without damaging the instrument.

■ **maximum power transfer**

a theorem that states that maximum power will be transferred from source to load when input impedance of the load equals the output impedance of the source.

■ **maximum response axis**

the MRa or acoustic axis of a transducer is defined as the direction in which the acoustic response has its maximum value.

■ **MCM**

Multi-Chip Module; an integrated circuit package which contains two or more interconnected chips.

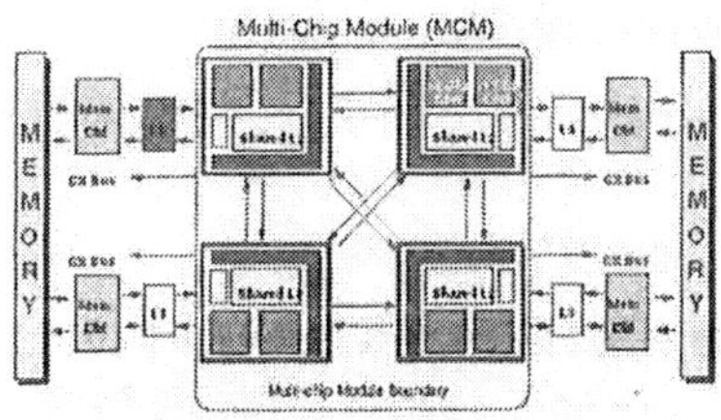

Multi-Chip Module

■ **MEMS**

Microelectromechanical Systems; integrated circuits which contain both electronic circuits and microscopic machines such as force transducers, valves, and motors.

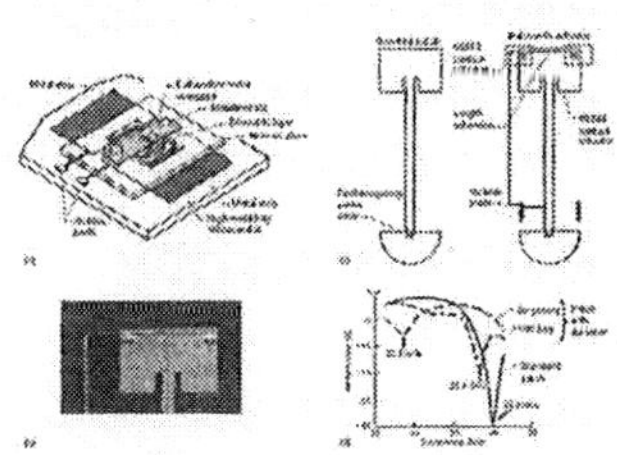

Microelectromechanical Systems

MESFET

MEtal-Semiconductor FET. a type of FET that has a Schottky barrier as the gate electrode rather than a semiconductor junction and I-V characteristics similar to a junction FET.

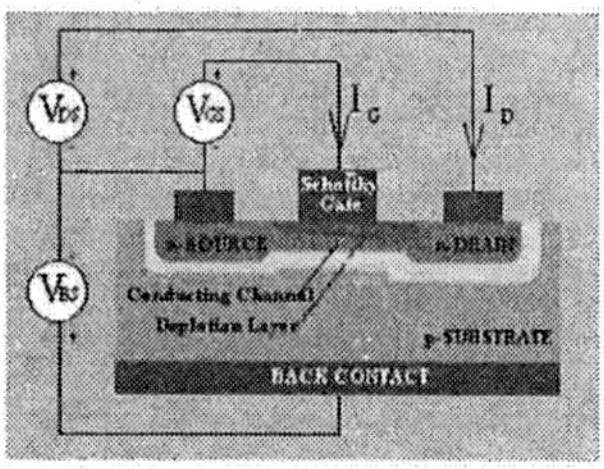

metal film resistor

a resistor in which a film of metal oxide or alloy is deposited on an insulating substrate.

METAL INSULATOR SEMICONDUCTOR FET (MISFET)

a type of FET in which the metal gate is insulated from the semiconductor material.

metal oxide field effect transistor

(MOSFET) a field effect transistor in which the insulating layer between the gate electrode and the channel is a metal oxide layer.

metal oxide resistor

a metal film resistor in which an oxide of metal (such as tin) is deposited as a film onto the substrate.

meter

any electrical or electronic measuring device. In the metric system, it is the unit of length equal to 39.37 inches.

mho

the unit of conductance, transconductance, or admittance; it is the word ohm spelled backwards.

micro switch

an analogue device used to input information into a system. Micro switches are useful because they require little movement to operate and are relatively sensitive Only small movement required to operate Input device .Physical sensor.

microcontroller

a microprocessor on a single integrated circuit intended to operate as an embedded system.

micromanipulator, micropositioner

an apparatus for holding and positioning a probe to contact a semiconductor wafer.

micrometer (micron)

one millionth (.000001) of a meter.(1 micron = 1 x 10^{-6} meters = 0.04 mils.).

microphone
electro acoustic transducer that converts sound energy into electric energy.

microwave
band of very short wavelength radio waves within the UHF, SHF and EHF bands.

midband gain
gain of an amplifier operating within its bandwidth.

mid-point bias
an amplifier biased at the centre of its dc load line.

Miller's theorem
a theorem that allows you to represent a feedback capacitor as equivalent input and output shunt capacitors.

minority carrier
the type of free charge carrier that does not predominate in an extrinsic semiconductor material. for n-type, these are holes and for p-type, these are electrons.

mismatch
term used to describe a difference between the output impedance of a source and the input impedance of a load. a mismatch prevents the maximum transfer of power from source to load.

mixed signal
describes a combination of analog and digital signals being generated, controlled or modified on the same chip.

mobile ion charge
charge present in the insulator or oxide of an MIS device, which under certain conditions will migrate, causing a shift in device characteristics. In silicon dioxide, sodium ($Na+$) and potassium ($K+$) are the predominant mobile ions.

mobility
a measure of the speed at which carriers move through doped semiconductor material per unit of electric field.

modal dispersion
the temporal spreading of a pulse in an optical wave guide caused by modal effects.

mode
a particular functioning arrangement, setup, or condition for laser operation, such as continuous emission, pulses, or grouped pulses. "Mode" also describes the cross-sectional shape of the beam.

mode field diameter
a measure of the core sise in single mode fiber.

mode filter
a device that removes optical power in higher order modes in fiber.

mode scrambler
a device that mixes optical power in fiber to achieve equal power distribution in all modes. Mode stripper a device that removes light in the cladding of an optical fiber.

mode shape
the resultant deflected shape of a rotor at a specific rotational speed to an applied forcing function. a three-dimensional presentation of rotor lateral deflection along the shaft axis.

modeling
the capture of the electrical and functional behavior of a circuit within a computer program which is then used to test the response of the circuit to various input data without actually fabricating the chip.

modem
a contraction of modulator/demodulator. a device that converts digital data to/from analog signals for transmission over analog lines.

modular angle encoder
angle encoder, consisting essentially of disk and scanning unit assemblies (rotor and stator) which are integrated into a machine

or a rotary table.

modulation
Process by which an information signal (audio for example) is used to modify some characteristic of a higher frequency wave known as a carrier (radio for example).

modulation index
the numerical value of the ratio of maximum FM deviation to the highest audio-modulating frequency (deviation ratio).

modulation, amplitude (AM)
the process where the amplitude of a signal is varied as a function of the instantaneous value of a another signal. The first signal is called the carrier, and the second signal is called the modulating signal. Amplitude modulation always produces a component at the carrier frequency, with components (sidebands) at the frequency of the carrier frequency plus minus the modulating signal.

modulation, frequency (FM)
the process where the frequency of the carrier is determined by the amplitude of the modulating signal. Frequency modulation produces a component at the carrier frequency, with adjacent components(sidebands) at frequencies around the carrier frequency related to the modulating signal. The carrier and sidebands are described by Bessel functions.

module
an interconnected subsystem of integrated circuit components, usually designed to act as a building block for more complex circuits.

moire principle
Photoelectric scanning method to produce periodic signals using two fine gratings which are closely positioned to each other and have approximately parallel and equally spaced lines.

molecule
smallest particle of a compound that still retains its characteristics. monochromatic light.
theoretically, light consisting of just one wavelength. Since no light is completely monochromatic, it usually consists of a very narrow band of wavelengths. Lasers provide the narrowest bands.

monopole
a single exposed radiator, usually requiring a ground plane to provide directivity (gain).

mono-stable
a system that has only one stable state. It can be made to change but always goes back to the original state. For example, a door with a door closer, It can be made to open but will always close. 555 timers can be set up as monostable devices .

monostable multivibrator
a multivibrator with one stable output state. When triggered, the circuit output will switch to the unstable state for a predetermined period of time and then return to the stable state. a timer.

monotonic (referring to the DAC)
a digital-to-analog converter (DAC) is said to be monotonic if, as the DAC code is increased, the analog output never decreases in value. a +/-1LSB DNL guarantees that a DAC is monotonic.

MOS
Metal-Oxide-Semiconductor; the fabrication technology which is

used to produce silicon field effect transistors in which the flow of current is controlled by the strength of the electric field between the gate and the conducting channel, which are separated by an insulating layer of silicon dioxide.

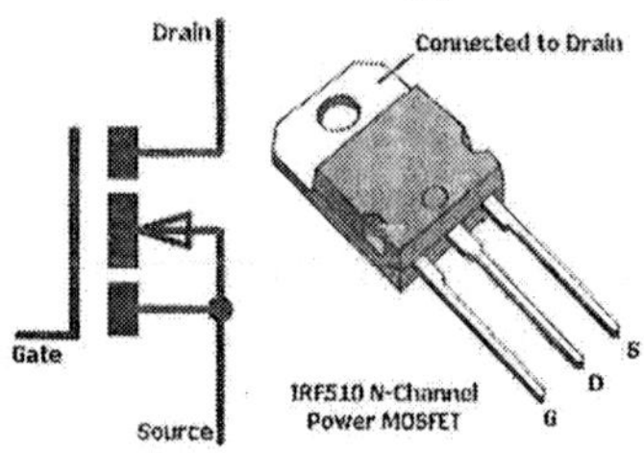

Mosfet Symbol & Packaging

■ MOS capacitor

a type of MIS capacitor in which the insulator is silicon dioxide and the semiconductor is typically silicon.

■ MOSFET

abbreviation for "metal oxide field effect transistor" also known as an "insulated gate field effect transistor). a field effect transistor in which the insulating layer betwen the gate electrode and the channel is a metal oxide layer.

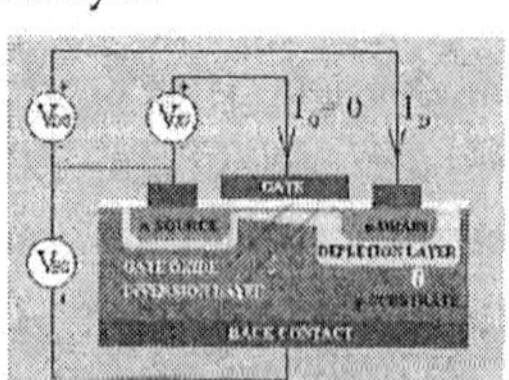

Crossection of a Metal-Oxide-Silicon Field-effect-Transistor

■ Most Significant Bit (MSB)

In the binary number scheme, the bit position in a group of bits given the largest weighted value. The MSB is typically the furthest-left bit in a grouping of data bits.

■ multimeter

electronic test equipment that can perform multiple tasks. Typically one capable of measuring voltage, current and resistance. More sophisticated modern digital multimeters also measure capacitance, inductance, current gain of transistors and/or anything else that can be measured electronically.

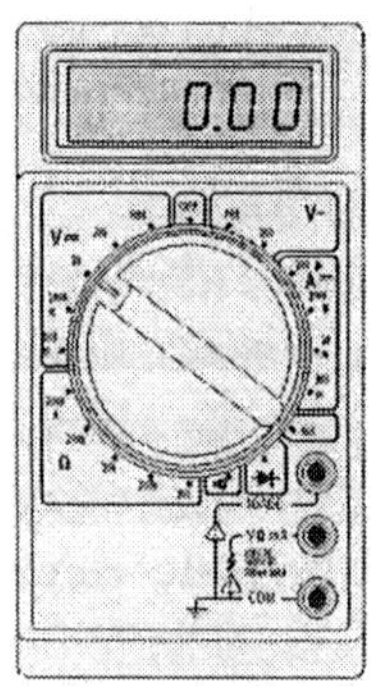

■ multimode fiber

a fiber with core diameter much larger than the wavelength of light transmitted that allows many modes of light to propagate. Commonly used with LED sources for lower speed, short distance links.

multiple frequency

at more than one frequency. This term is applied to a capacitance meter that offers a choice of discrete test signal frequencies.

multiplex

connecting one instrument to multiple devices under test or multiple instruments to one device under test .

multiplexer

also commonly referred to as a "MUX." a device used to combine multiple parallel information streams into a single channel. There are a number of different types of multiplexers: Frequency Division Multiplexing (FDM), Time Division Multiplexing (TDM), and Statistical Time Division Multiplexing (STDM).

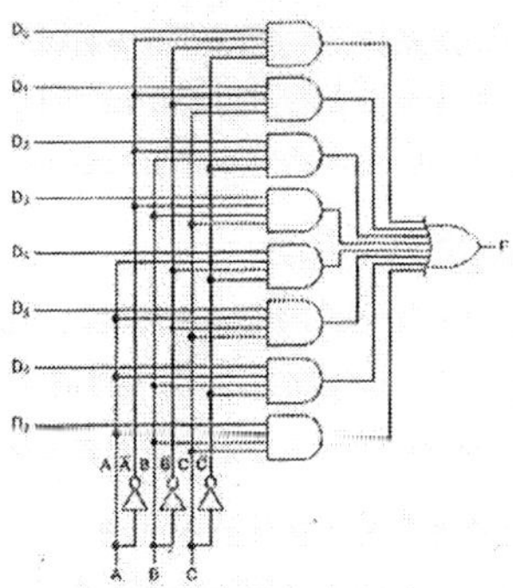

Multiplexer Schematic Diagram

multiplier resistor

resistor connected in series with a moving coil meter movement to extend the voltage ranges.

multiplying DAC

a digital-to-analog converter (DAC) that allows an AC signal to be applied to the reference input pin. This allows the DAC to be used as a digital attenuator by feeding the signal of interest into the reference pin and using the DAC codes to scale it.

multisegmant display

device made of several light emitting diodes arranged in a numeric or alphanumeric pattern. By lighting selected segments numeric or alphabet characters can be displayed.

multiturn rotary encoder

absolute rotary encoder which determines the angular position of the shaft and the number of shaft rotations.

multivibrator

a class of circuits designed to produce square waves or pulses. Astable multivibrators produce continuous pulses without an external stimulus or trigger. Monostable multivibrators produce a single pulse for some predetermined period of time only when triggered. Bistable multivibrators produce a DC output which is stable in either one of two states. Either high or low. an external stimulus or trigger is required for the bistable circuit to change states, either high to low or low to high.

■ MXN crosspoint (or matrix) switch

a matrix of switches configured with M inputs and N outputs. Under digital control, any output may be connected to one of the inputs.

■ NAND gate

gate whose output is 0 when both inputs are 1, and 1 in all other cases.

■ nanometer

a unit of length in the International System of Units (SI) equal to one billionth of a meter (10^{-9} meter). Once called a millimicron, it is used to represent wavelength. Abbreviated "nm."

■ nanovoltmeter

a sensitive DC voltmeter (typically one decade more sensitive than a digital millimetre) with a low thermal input connection.

■ natural frequency

the frequency of free vibration of a system. The frequency at which an undamped system with a single degree of freedom will oscillate upon momentary displacement from its rest position.

■ Nd:GLASS LASER

a solid-state laser of Neodymium: glass offering high power or short pulses, or both, for specific industrial applications.

■ near field imaging

a solid-state laser imaging technique offering control of spot sise and hole geometry, adjustable working distance, uniform energy distribution, and easily produced range of spot sizes.

■ negative charge

a charge that has more electrons than protons.

■ negative feedback

a feedback signal 180° out of phase with an amplifier input signal. Used to increase amplifier stability, bandwidth and input impedance. Also reduces distortion.

■ negative ion

an atom having a greater number of electrons in orbit than there are protons in the nucleus.

■ negative resistance

a resistance such that when the current through it increases the voltage drop across the resistance decreases.

■ **negative supply**

any supply which has a voltage *less* than zero VOLTS. Some circuits, particularly OPERATIONAL AMPLIFIERS, need a negative supply to work reliably. This can be obtained using special equipment, or by using a VOLTAGE DIVIDER set to split the voltage in half and connecting the halfway voltage to the circuits EARTH point. The first method is preferred.

■ **negative temperature coefficient**

a term used to describe a component whose resistance or capacitance decreases when temperature increases.

■ **NEMA**

national Electrical Manufacturers' Association, a group which defines and recommends safety standards for electrical equipment.

■ **net list**

a computer file listing all of the components together with the details of how they are interconnected to form the overall circuit.

■ **network**

a system of cables, hardware and equipment used for communications.

■ **network analysis**

the process of creating a model of the transfer or impedance characteristics of a linear network through stimulus-response testing over the frequency range of interest.

■ **noutral**

a terminal, point or object with balanced charges. Neither positive or negative.

■ **neutral atom**

an atom in which the number of negative charges (electrons in orbit) is equal to the number of positive charges (protons in the nucleus).

■ **neutral wire**

the conductor of a polyphase circuit or a single-phase three wire circuit that is intended to have a ground potential. The potential difference between the neutral and each of the other conductors are approximately equal in magnitude and equally spaced in phase.

■ **node**

junction or branch point in a circuit.

■ **noise**

unwanted electromagnetic radiation within an electrical or mechanical system. an operational amplifier circuit having no phase inversion between the input and output.

■ **non-equilibrium**

the condition in a semiconductor device or material when there is still a tendency for its macroscopic properties to change with time. In other words, there still remains some settling to occur before equilibrium is achieved. Devices in which current always flows, like diodes and bipolar transistors, do not ever reach equilibrium.

■ **non-inverting input**

the terminal on an operational amplifier that is identified by a plus sign.

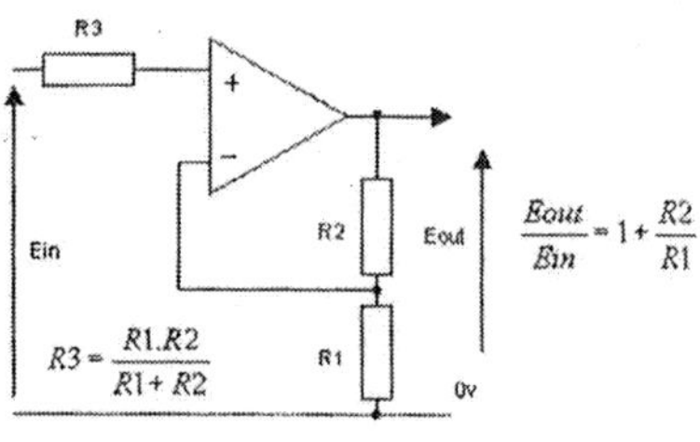

■ **nonlinear distortion**

a type of waveform distortion affecting the signal amplitudes, where one alternation of ac may have its amplitude changed with respect to the other.

■ **non-linear scale**

a scale in which the divisions are not equally spaced.

■ **non-linearity**

the maximum deviation of the sensor outputs, obtained for increasing input, from a line drawn between zero output and the rated output. Expressed as a percentage of rated output.

■ **NOR Gate**

a logic circuit used in computers and industrial circuitry which combines the NOT function with the OR.

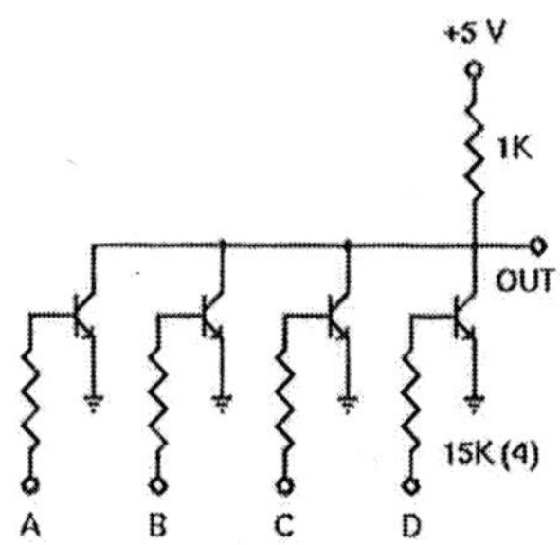

4 Input Nor Gate

■ **Normal Mode Rejection Ratio (NMRR)**

the ability of an instrument to reject interference (usually of line frequency) across its input terminals. Usually expressed in decibels at a frequency.

■ **normal mode voltage**

a voltage applied between the input high and input low terminals of an instrument.

■ **normal closed**

designation which states that the

contacts of a switch or relay are closed or connected when at rest. When activated, the contacts open or separated.

normally open
designation which states that the contacts of a switch or relay are normally open or not connected. When activated the contacts close or become connected.

north pole
pole of a magnet out of which magnetic lines of force are assumed to originate.

Norton's theorem
any network of voltage sources and resistors can be replace by a single current source in parallel with a single resistor.

NOT Gate
1. a logic circuit in which the output signal has a polarity opposite to the input signal.
2. Logic element that acts as an inverter.

notch filter
a filter which blocks a narrow band of frequencies and passes all frequencies above and below the band.

NPN transistor
a bipolar junction transistor in which a p-type base element is sandwiched between an n-type emitter and an n-type collector.

interrupt
a signal to the CPU indicating that the board detected the occurrence of a specified condition or event.

NTSC
National Television Standards Committee. NTSC is the video standard used in North America and some other parts of the world to get video into your home and to record onto videotape. The NTSC system uses 525 lines per frame, a 30-frame per second update rate, and the YIQ color space. Modern NTSC encoders and decoders may also use the YUV color space.

***n*-type semiconductor**
a semiconductor compound formed by doping an intrinsic semiconductor with a pentavalent element. an n-type material contains an excess of conduction band electrons.

nucleus
Core of an atom. The nucleus contains both positive (protons) and neutral (neutrons) subatomic particles.

numerical aperture (NA)
a measure of the light acceptance angle of the fiber.

■ Nyquist Criterion

requirement that a sampled system needsto be sampled at a frequency greater than twice the bandwidthof the signal to be sampled.

■ Nyquist frequency

in sampling theory, the Nyquist frequency is the maximum frequency that can be applied to the analog-to-digital converter (ADC) input with no aliasing effects. The Nyquist frequency is defined as the "sampling frequency/2." *See also* aliasing.

■ Nyquist Plot

a plot of real versus imaginary spectralcomponents that is often used in servo analysis. Should not beconfused with a polar plot of amplitude and phase of 1x vibration.

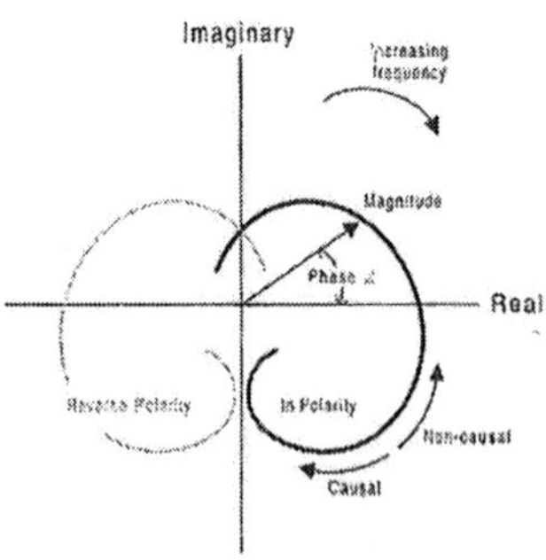

■ object

the subject matter or figure imaged by, or seen through, an optical system.

■ Object Linking And Embedding (OLE)

a Microsoft Windows standard mechanism for embedding one program within another. For example, an Excel spreadsheet can be pasted into a Visual Basic program. If a file is linked to an OLE control, the data stored in that file is displayed in the OLE control.

■ octave

interval between two sounds whose fundamental frequencies differ by a ratio of 2 to 1. 440 Hz. is one octave above 220 Hz.

■ offset binary coding

For bipolar signals, offset binary is a digital coding scheme in which the most negative value is represented by all zeros (00000000) and the most positive value is represented by all ones (11111111).

■ offset current

a current that comes from a switching card even though no user signals are applied. It comes mostly from the finite coil to contact impedance. It is also generated by triboelectric, piezoelectric, and electrochemical effects present on the card.

off-set dipole

a vertically aligned dipole off-set (normally a 1/4 wave of 1/2 wavelength of the operating frequency) from a large vertical element (the mast).

offset error (bipolar)

when using a bipolar converter, the ideal midscale transition occurs at AGND -0.5LSB. Bipolar offset error is the measured deviation from this ideal value. *See also* offset error (unipolar).

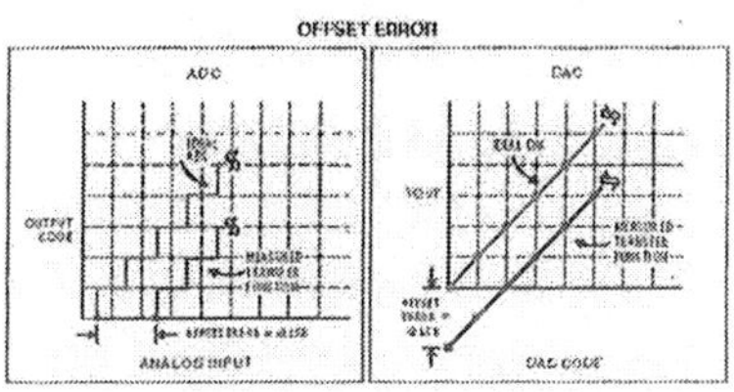

offset error (unipolar)

For an ideal converter, the first transition occurs at 1/2LSB, above zero. Offset error is the amount of deviation between the measured first transition point and the ideal point. For an analog-to-digital converter (ADC), 0V is applied to the ADC input and then the analog input voltage is increased until the first transition occurs. This voltage can be converted to LSBs by dividing it by the LSB step size, which is VREF/2N. For a digital-to-analog converter (DAC), offset error is determined by loading a code of all zeros into the DAC and measuring the analog output voltage. Offset error is easily calibrated out in software or with hardware.

offset null

an op amp control pin used to eliminate the effects of internal component voltages on the output of the device.

ohm

unit of resistance symbolised by the Greek capital letter omega (?).

ohm

the unit of resistance (the voltage value divided by the current value).

Ohm's Law

a basic law of electricity establishing the mathematical relationships between current, voltage, and resistance values.

ohmmeter

device used to measure electrical resistance.

omni directional

sending or receiving sound waves in or from any direction. 360 degrees receiving capability.

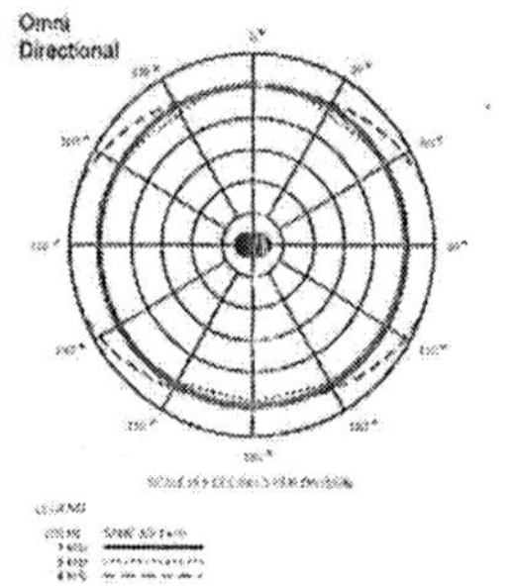

Omni Directional Pickup Pattern

■ one-shot
Monostable multivibrator

■ op-amp
abbreviation for operational amplifier.

■ open circuit voltage
the OCV is the level of the electrical output per one micropascal of acoustic input.

■ open loop buffer amplifier
a unity gain (gain = 1) amplifier not using a feedback network (loop). Thus it's "open loop"

■ open loop gain
gain of an amplifier when no feedback is present.

■ open loop mode
an amplifier circuit having no means of comparing the output with the input. (No feedback.).

■ operating point
that point in the characteristics of a tube or transistor around which the signal voltages vary.

■ operating system
Base-level software that organises the computer's resources and capabilities, runs application programs, interacts with users, and communicates with installed and peripheral devices. Popular operating systems include DOS, Windows, OS/2, and UNIX.

■ operation modes
refers to single, synchronous, interrupt, DMA, and on-board memory (OBM) operations.

■ operational amplifier
a high gain DC amplifier that has a high input impedance and a low output impedance. Op-amps are the most basic type of linear integrated circuits.

■ operational transconductance amplifier (OTA)
a differential amplifier that converts the difference between the input voltages to a proportional output current.

■ optical amplifier
a device that amplifies light without converting it to an electrical signal.

■ optical density
protection factor provided by a

filter (such as used in eyewear, viewing windows, etc.) at a specific wavelength. Each unit of OD represents a 10x increase in protection.

optical fiber
an optical waveguide, comprised of a light carrying core and cladding which traps light in the core.

optical loss test set (OLTS)
an measurement instrument for optical loss that includes both a meter and source.

optical power
the amount of radiant energy per unit time, expressed in linear units of Watts or on a logarithmic scale, in dBm (where 0 dB = 1 mW) or dB* (where 0 dB*=1 microWatt).

optical pumping
exciting the lasing medium by the application of light rather than electrical discharge from anode and cathode.

optical return loss, back reflection
Light reflected from the cleaved or polished end of a fiber caused by the difference of refractive indices of air and glass. Typically 4% of the incident light. Expressed in dB relative to incident power.

optical return loss, back reflection
Light reflected from the cleaved or polished end of a fiber caused by the difference of refractive indices of air and glass. Typically 4% of the incident light. Expressed in dB relative to incident power.

optical switch
a device that routes an optical signal from one or more input ports to one or more output ports.

optical time domain reflectometer (OTDR)
an instruments that used backscattered light to find faults in optical fiber and infer loss.

opto isolator
an optical switch contains a light-emitting diode (LED) and a phototransistor. The LED emits light, and the phototransistor senses light. The phototransistor conducts when light falls on it.

opto electronics
integrated circuits and systems which combine optical elements (e. g. lasers and photo detectors) with microelectronic circuits, often as part of a single chip.

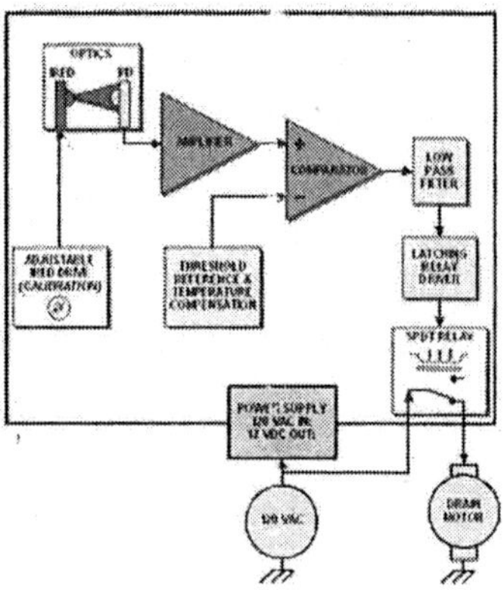

Turbidity Sensing

■ **OR gate**

a logic gate circuit in which an output is procured for the application of one or more signals to the input terminals.

■ **oscillator**

an electronic circuit that produces a continuous output waveform with only DC applied.

■ **oscillator-demodulator**

a signal conditioning device that sends a radio frequency signal to an eddy-current displacement probe, demodulates the probe output, and provides output signals proportional to both the average and dynamic gap distances.

■ **oscilloscope**

an instrument used to display a signal graphically. Shows signal amplitude, period and waveshape in addition to any DC voltage present. a multiple trace oscilloscope can show two or more waveforms at the same time for phase comparison and timing measurements.

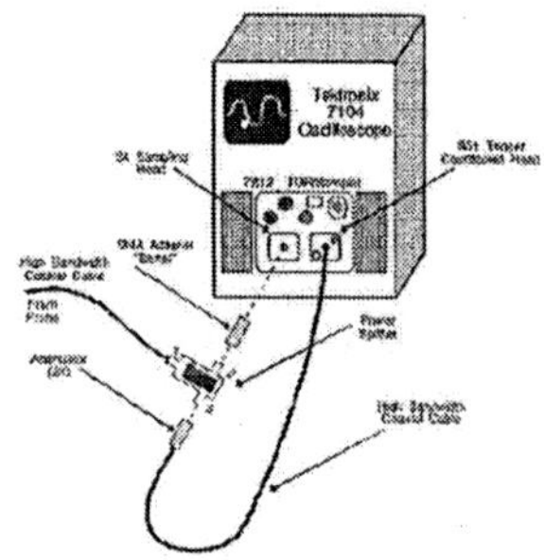

■ **out of phase**

when the maximum and minimum points of two or more waveshpes do not occur at the same time.

■ **output impedance**

the equivalent impedance represented in series with an analog output. The smaller the impedance the better one can drive heavier loads.

■ **output limiting buffer**

a buffer with external pins (VH and VL) that define the maximum positive and negative output swing. Connecting reference voltages to these pins "limits" the output swing to this voltage range.

■ **output power**
amount of power a component, circuit or system can deliver to a load.

■ **output voltage range**
the maximum and minimum signal which an analog output can provide and stay within a recommended operating range (e.g., no overdriving occurs).

■ **overcoupling**
the degree of coupling beyond the critical point in resonant circuits which produces a dip at the resonant-frequency point on the response curve.

■ **overfilled launch**
a condition for launching light into the fiber where the incoming light has a spot sise and Na larger than accepted by the fiber, filling all modes in the fiber.

■ **overlap capacitances**
a CMOS device parameter. Source/drain to gate capacitance per unit length.

■ **overload**
condition that occurs when the load is greater than the system was designed to handle. (Load resistance too small, load current too high.) Overload results in waveform distortion and/or overheating.

■ **overload protection**
protective device such as a fuse or circuit breaker that automatically disconnects a load when current exceeds a predetermined value.

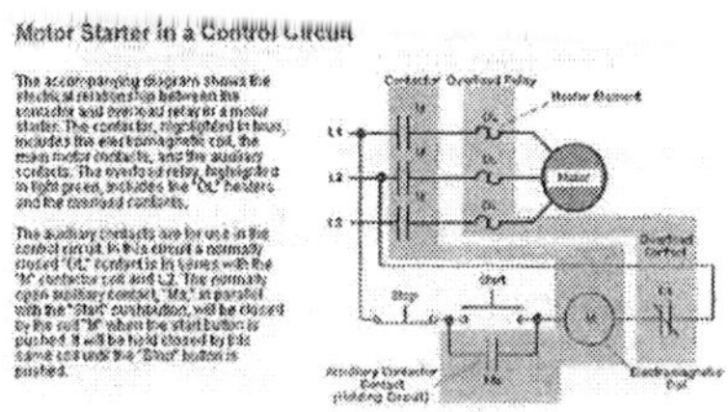

Overload Protection in a Motor Circuit

■ **overload protection**
a circuit that protects the instrument against excessive current at the input terminals.

■ **overload-safe**
the maximum input in excess of the rated input that can be applied along the principal axis without producing a permanent change in the sensor specification. Expressed as a percentage of the rated measurand.

■ **overmodulation**
the modulation of an r-f carrier in excess of what is defined as 100%.

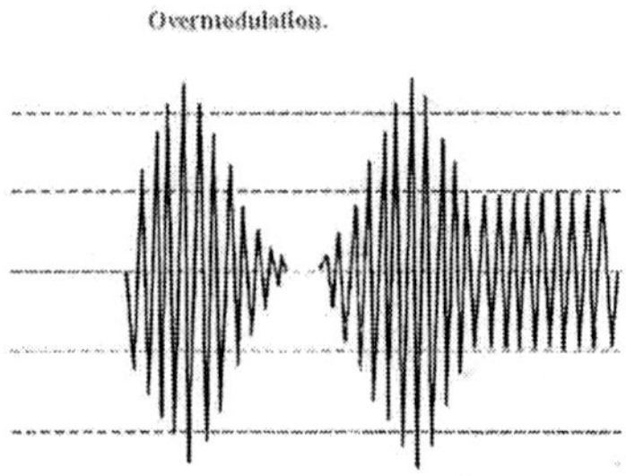

Signal Loss Due to Overmodulation

■ **oversampling**

sampling the analog-to-digital converter (ADC) input at a much higher frequency than that of Nyquist. This technique provides a processing gain by effectively reducing the noise floor of the converter. a 2X increase in over sampling rate theoretically improves signal-to-noise by 3dB. *See also* Nyquist.

■ **overtone**

a harmonic frequency related to the fundamental. See Harmonic.

■ **overvoltage clamp**

a device to restrict input voltages to acceptable limits.

■ **oxidation**

the growth of a layer of silicon dioxide on a silicon surface.

■ **oxide integrity**

the quality of the oxide. Oxide integrity testing often includes C-V, along with breakdown and reliability tests.

■ **oxide thicknesses. (parametric testing)**

a CMOS device parameter. Equivalent thickness of thin, field, and inter-level dielectrics.

■ **oxide trapped charge**

Charge trapped in the insulator of an MIS device that cannot move to the semiconductor material. Oxide trapped charge may anneal during a bias temperature stress cycle. Oxide trapped charge is located near the Si-SiO2 interface.

■ **oxide**

a casual term for silicon dioxide (otherwise known as glass). This material is used in semiconductor processing as an insulator and to cover and protect or "passivate" the device.

■ **oxidizer**

a chemical other than a blasting agent or explosive that initiates or promotes combustion by spontaneously evolving oxygen either at room temperature or under slight heating. These chemicals can react with organic material or intensify fires.

■ **oxygen assist**

in certain cutting operations, coaxial oxygen initiates an exothermic reaction to enhance the cutting rate for thick metals; in other words, oxygen actually does the cutting, with the reaction being maintained by the laser beam.

paced mode

the sampling of each analog input channel at regularly spaced intervals. This interval is determined by the conversion rate of the analog-to-digital converter and the number of channels sampled. In this mode, the conversion rate equals the pacer clock rate.

pacer clock

an on-board or external clock that paces or times events such as analog-to-digital conversions, digital-to-analog conversions, data sampling, interrupt generation, digital I/O transfers, etc.

package

the plastic or ceramic structure that encloses a chip and facilitates connections to the chip through a set of pins, which are usually soldered to a printed circuit board. The package may be first inserted into a socket to permit easy removal.

pad

an attenuator circuit employed for coupling the output impedance of one unit or circuit to the input impedance of another.

PAL

PAL stands for Phase Alternation Line. PAL is to Europe as NTSC is to North America. In other words, PAL is the video standard used in Europe and a few other countries. PAL uses 25 lines per frame, while NTSC has 525 lines. Therefore, PAL has higher vertical resolution. The frame rate of NTSC is 30 frames per second, while for PAL it is 25. This means the update rate for NTSC is higher and, therefore, there is more flicker with PAL. PAL uses the YUV color space, while NTSC uses YIQ or YUV.

paper capacitor

Fixed capacitor using oiled or waxed paper as a dielectric.

parallel

circuit having two or more paths for current flow. Also called shunt.

Parallel

resistors, cells, switches, and batteries etc can be connected in parallel as in this diagram.

parallel resonant circuit

circuit having an inductor and a capacitor in parallel with one another. Circuit offers a high impedance at resonant frequency. Sometimes called a "tank circuit."

■ pasive filter
a filter that contains only passive or non amplifying components.

■ pass band
the range of frequencies that will be passed and amplified by a tuned amplifier. Also the range of frequencies passed by a band pass filter.

■ pass transistors
transistors used to generate a stable supply of voltage from an unstable input.

■ passivation
protection of the junctions and surfaces of solid state electronic components and ICs from harmful environments; most commonly achieved by forming a layer of silicon dioxide on the silicon chip.

■ passive component
component that does not amplify a signal. Resistors and capacitors are examples.

■ passive system
system that emits no energy. It only receives. It does not transmit or reveal its position.

■ path isolation
on a matrix switching card, the isolation from signal high and low of one path to signal high and low of any other path. Specified as resistance and capacitance..

■ path resistance
on a matrix switching card, the resistance per conductor of a closed path, including the tape resistance and connector terminal resistance.

■ path
one of many signal paths on a matrix switching card. a path is established by the actuation of a relay at a row and column crosspoint.

■ PC
abbreviation for Personal Computer.

■ PCB
Printed Circuit Board; an insulating card with one or more layers of copper interconnections onto which are soldered integrated circuit packages and other electronic components. PCBs are usually plugged into sockets that may be interconnected through a wire bus.

Printed Circuit Board

■ **peak**
maximum or highest amplitude level.

■ **peak inverse voltage**
(PIV) The maximum rated value of a AC voltage acting in the direction opposite to that in which a device is designed to pass current.

■ **peak to peak**
difference between the maximum positive and maximum negative values of an AC waveform.

■ **pentavalent element**
element whose atoms have five valence electrons. Used in doping intrinsic silicon or germanium to produce n-type semiconductor material. Most commonly used pentavalent materials are arsenic and phosphorus.

■ **percent of regulation**
the change in output voltage that occurs between no-load and full-load in a DC voltage source. Dividing this change by the full-load value and multiplying the result by 100 gives percent regulation.

■ **percent of ripple**
the ratio of the effective rms value of ripple voltage to the average value of the total voltage. Expressed as a percentage.

■ **phase**
angular relationship between two waves.

■ **phase angle**
Phase difference between two or more waves, normally expressed in degrees.

■ **phase evaluation**
method of determining position by detecting the phase between alternating
voltages having a slight variation in frequency.

■ **phase grating scale**
scale with step grating which diffracts the transmitted or reflected light into 2 or more orders.

■ **phase inversion**
a circuit for obtaining a reversal of the phase of a signal.

■ **phase shift**
Change in phase of a waveform

between two points, expressed as degrees of lead or lag.

phase shift oscillator

an oscillator that uses three RC networks in its feedback path to produce the 180° phase shift required for oscillation.

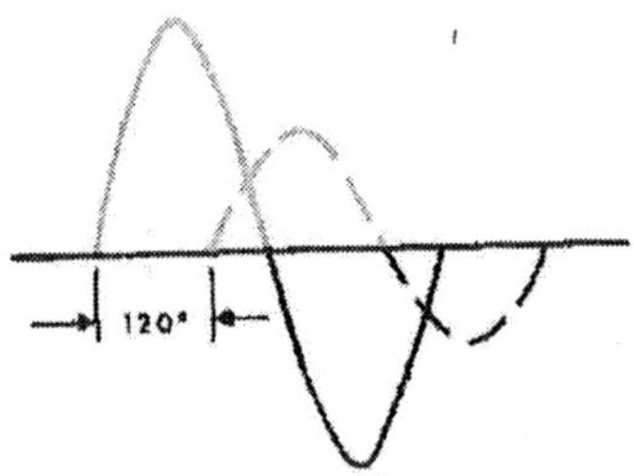

(B) PHASE-SHIFT NETWORK C2 AND R2

phase splitter

Circuit that takes a single input signal and produces two output signals that are 180° apart in phase.

phonograph

piece of equipment used to reproduce sound stored on a disk called a phonograph record.

phosphor

luminescent material applied to the inner face of a cathode ray tube that when bombarded with electrons will emit light of various colours.

photoacoustic effects

arises with the use of very short-duration high-energy laser pulses, at pulse durations typically below 10 microseconds. Significant amounts of energy are absorbed and a rapid expansion occurs in the tissue, generating an acoustic shock wave that causes mechanical disruption to cellular structures.

photocell

a device that changes resistance when light falls on it. Similar to a light dependent resistor or LDR. This cell does not produce an output current -.

photochemical effects

effects that occur from long exposure durations at incident power levels insufficient to cause damaging photothermal effects. It is an energy dependent process (a function of the total quantity of radiation absorbed rather than its rate of absorption).

photoconduction

a process by which the conductance of a material is change by incident electromagnetic radiation in the visible light spectrum.

photoconductive cell

material whose resistance decreases or conductance increases when exposed to light.

■ photoconductor

this is a special type of RESISTOR and is sometimes called an LDR. Photoconductors are made so that their resistance decreases as the level of light falling on them increases.

■ photodetector

Component used to detect or sense light.

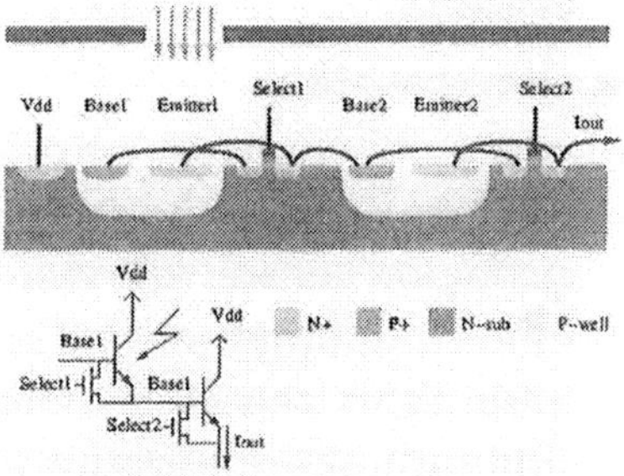

Schematic of Photodetector

■ photodiode

a semiconductor diode that changes its electrical characteristics in response to illumination.

■ photoelectric cell

the term generally applied to a unit whose electric characteristics are sensitive to the intensity of incident light.

■ photo-mask

the photographic film (glass plate) which contains the masking pattern required for a particular wafer processing step.

■ photometer

an instrument which measures luminous intensity.

■ photon

in quantum theory, the elemental unit of light, having both wave and particle behavior. It has motion, but no mass or charge.

■ photoresist

light-sensitive material that is used to form selectively protective masking layers during wafer processing. a photo-mask is used to define the areas of photoresist exposed to light, which are subsequently removed by a solvent, leaving the required pattern on the wafer surface. The entire process is called photolithography.

■ photoresistor

also known as a photoconductive cell or light dependent resistor. (LDR) a device whose resistance decreases with exposure to light.

■ photothermal effects

the damage mechanism for acute laser injury (i.e. for injury immediately following exposure). The radiation incident at the surface is absorbed in the underlying tissue, increasing the temperature of the tissue to the level at which damage can occur, and laser burns result. It is a power dependent process (a function of the RATE at which energy is absorbed rather than the total quantity of energy involved).

■ **phototransistor**

this type of TRANSISTOR has only two pins, the COLLECTOR and EMITTER. Because they have no BASE pin, they can sometimes be mistaken for a normal DIODE or an LED. The amount of light falling on them acts instead of the base current and once a certain level is reached, the transistor is switched on.

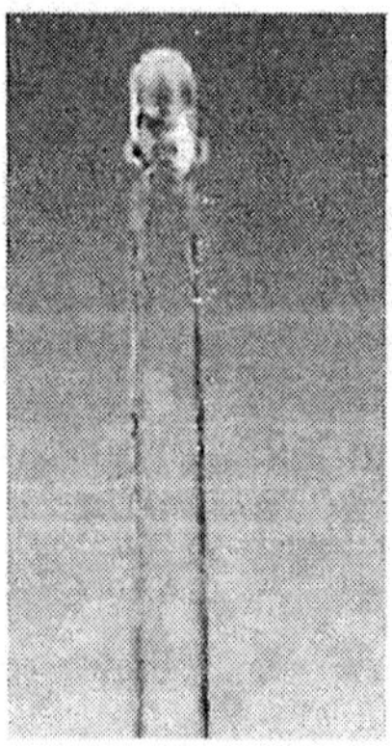

Infrared Phototransistor

■ **photovoltaic cell**

component commonly called a solar cell used to convert light energy into electrical energy.

■ **picoammeter**

a measuring instrument that is similar in function to the ammeter of an electrometer. General differences include as low or lower voltage burden, faster readings, and less sensitivity.

■ **pierce oscillator**

a variation of the Colpitts oscillator. This oscillator uses a quartz crystal in place of the inductor found in the Colpitts oscillator feedback network. The crystal maintains a highly stable output frequency.

■ **piezo sounder**

small crystal element that can emit very high sound levels while requiring low current. Also called a Piezo diaphragm as it has no active circuit connected to the diaphragm. a piezo buzzer has an active circuit connected to the diaphragm to emit sound when a DC voltage is applied.

■ **piezoelectric**

any material, which provides a conversion between mechanical and electrical energy. For a piezoelectric crystal, if mechanical stresses are applied on two opposite faces, electrical charges appear on some other pair of faces.

■ **piezo-electric ceramic**

a material made of crystalline substance, which creates charges of electricity by the application of pressure and vice versa.

■ piezoelectric crystal

crystal material that will generate a voltage when mechanical pressure is applied and conversely will undergo mechanical stress when subjected to a voltage.

■ piezoelectric effect

the production of a voltage between opposite sides of a piezoelectric crystal as a result of pressure or twisting. Also the reverse effect which the application of a voltage to opposite sides causes a deformation to occur at the frequency of the applied voltage. (Converts mechanical energy into electrical energy and electrical energy into mechanical energy.).

■ pigtail

a short length of fiber attached to a fiber optic component such as a laser or coupler.

■ pin driver

a very high performance 2 channel MUX used to switch between 2 voltages at a very high rate, and drive a test "pin" on automated test equipment (ATE).

■ pinch-off region

a region on the characteristic curve of a FET in which the gate bias causes the depletion region to extend completely across the channel.

■ plasma

in laser welding, a metal vapor that forms above the spot where the beam reacts with the metal surface. Also used to describe the laser tube (plasma tube, discharge tube) which contains the completely ionized gas in certain lasers.

■ plastic film capacitor

Capacitor in which alternate layers of aluminium foil are separated by thin films of plastic dielectric.

■ plate

Conductive electrode in either a capacitor or battery. In vacuum tube technology, it is the name given to the anode.

■ plate modulation

the introduction of the modulating signal into the anode circuit of the r-f carrier amplifier.

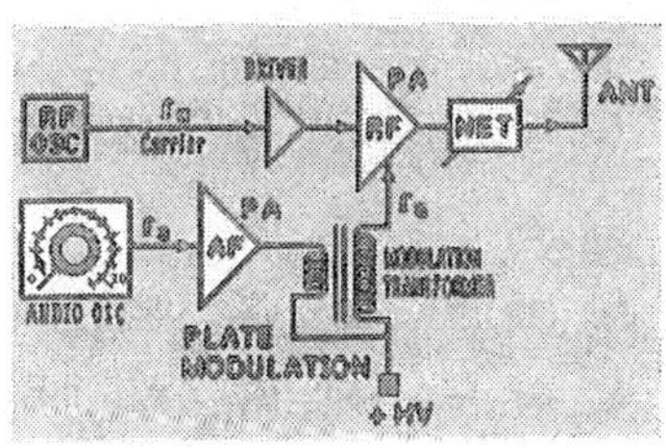

Block Diagram of Plate Modulation

■ plate resistance

the dynamic (a-c signal) resistance of a tube. It is the value obtained when a small change in plate voltage is divided by a small change in plate current, for a constant bias value.

■ **plug**

movable connector that is normally connected into a socket or jack.

■ **PN junction**

the interface between n-type and p-type semiconductor material. This structure forms a pn junction diode.

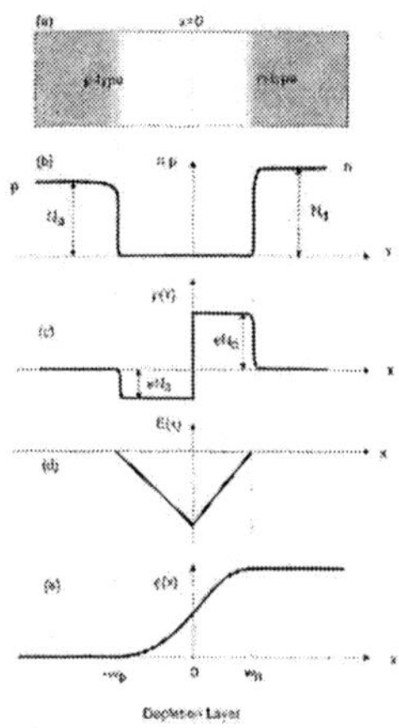

PN Junction with Energy Diagram and Depletion Layer

■ **PNP**

this refers to TRANSISTORS, which usually have three pins. The three letters correspond to the pins, with the middle one being the BASE. The 'N' in this case means that the base must be 0.6 volts *lower* than then EMITTER before the transistor will switch on. It is the opposite of NPN.

■ **pointing errors**

beam movement and divergence, often preventable by using short path-lengths.

■ **polar coordinates**

either of two numbers that locate a point in a plane by its distance from a fixed point and the angle this line makes with a fixed line.

■ **polar plot**

Polar coordinate representation of the locus of the 1x vector at a specific lateral shaft location with the shaft rotational speed as a parameter.

■ **polarised**

a component or plug etc that must be fitted around a certain way. eg. Thick and thin pins spaced apart so that the plug cannot be inserted around the wrong way.

■ **polarity**

term used to describe positive and negative charges.

■ **polarisation**

restriction of the vibrations of the electromagnetic field to a single plane, rather than the innumerable planes rotating about the vector axis. This prevents optical losses at interfaces between the lasing medium and optical elements. Various forms of polarization include random, linear (plane), vertical, horizontal, elliptical, and circular. Of two polarization components (so-called), S and P, the P component has zero losses at Brewster's angle l = wavelength E = electric vector H = magnetic vector.

pole
in an active filter, a single RC circuit. a one pole filter has one capacitor and one resistor. a two pole filter has two RC circuits and so on.

population inversion
when more molecules (atoms, ions) in a laser are in a metastable state than in the ground state (a situation needed for sustaining a high rate of stimulated emissions), a "population inversion" is said to exist. Without a population inversion, there can be no lasing action.

positive
polarity of point that attracts electrons as opposed to negative which supplies electrons.

positive charge
a charge that exists in a body that has fewer electrons than protons.

positive feedback
a feedback signal that is in phase with an amplifier input signal. Positive feedback is necessary for oscillation to occur.

positive ground
a system whereby the positive terminal of the source is connected to the system's conducting chassis.

positive ion
atom that has lost one or more valence electrons resulting in a net positive charge.

positive supply
any supply which has a voltage *greater* than zero VOLTS. Normally used to refer to the original power source.

potential barrier
the internal resistance of a transistor related to the internal potential existing between NP junctions.

potential difference
Voltage difference between two points which will cause current to flow in a closed circuit.

potential energy
energy that has potential to do work because of its position relative to others.

potentiometer
a variable resistor with three terminals. Mechanical turning of a shaft can be used to produce variable resistance and potential. Example: a volume control is usually a potentiometer.

power
amount of energy converted by a circuit or component in a unit of time, normally seconds. Measured in units of watts. (joules/second).

power amplifier
an amplifier designed to deliver maximum power output to a load. Example: In an audio system, it is the power amplifier that drives the loudspeaker.

power density
the amount of radiant energy concentrated on a surface.

power derating factor
a transistor rating that tells how much the maximum allowable value of PD decreased for each 1°C rise in ambient temperature.

power dissipation
amount of heat energy generated by a device in one second when current flows through it.

power factor
the cosine of the phase angle between voltage and current. Also R/Z.

power loss
ratio of power absorbed to power delivered.

power management subsystem
responsible for overall power consumption and supply of other subsystems in mobile phones, primarily affecting the total talk and standby times of a mobile phone.

power meter
an accessory used to monitor laser beam power at the rear reflector, tune the beam for optimum power, or monitor power delivered to the work station.

power on reset
describes the process of initialising a system in a known state.

power ramping
a controlled change in the power level of a laser beam, either linearly, as up a ramp, across, and down again—or in several discrete steps. Useful for smooth completion of circular welds and for preventing fractures from rapid cooling.

power supply rejection
the measure of how much a variation in power supply voltage couples through to an output signal of an analog circuit. If the PSR is divided by the system gain (hence input-referred), the term becomes PSRR, or power-supply rejection ratio.

preamplifier
an additional stage of amplification preceding a receiver or audio amplifier to increase signal amplitudes above a prescribed level.

precision op amp
an op amp with very high open loop gain and common mode/ power supply rejection ratios, and very low offset voltage and offset current. It is ideal for accurately amplifying signals, while introducing minimal error .

pre-emphasis
the gradual increase in amplitude of higher-frequency signals in frequency modulation practices for increasing signal-to-noise ratios.

preform
the large diameter glass rod from which fiber is drawn.

preload, bearing
the dimensionless quantity that is typically expressed as a number from zero to one where a preload of zeroindicates no bearing load upon the shaft, and one indicates themaximum preload (i.e., line contact between shaft and bearing).

preload, external
any of several mechanisms that can externally load a bearing. This includes "soft" preloads suchas process fluids or gravitational forces as well as "hard"preloads from gear contact forces, misalignment, rubs, etc.

primary
first winding of a transformer. Winding that is connected to the source as opposed to secondary which is a winding connected to a load.

primary cell
cell that produces electrical energy through an internal electrochemical action. Once discharged a primary cell cannot be reused.

printed circuit board
insulating board containing conductive tracks for circuit connections.

probe tip
the tip of a probe for contacting a semiconductor device. The probe looks like a fine needle that

is electrically connected to instruments. The probe is mechanically supported by a micropositioner.

■ prober

an apparatus for contacting devices on semiconductor wafers. This includes a plate on which the micropositioner rests, a chuck on which the wafer rests, and a microscope for viewing small devices while positioning probes.

■ programmable gain buffer

an amplifier that can be used in a fixed number of user-selectable gains (typically +/-1 and 2). The feedback network is on the IC, thus saving board space.

■ programmable UJT

unijunction transistor with a variable intrinsic stand-off ratio.

■ programmed I/O

a standard method of accessing an I/O device-the CPU reads each byte of data from or writes each byte of data to the device.

■ projector

a projector converts the energy from a power amplifier (generator) into an acoustic pressure output. Projectors are usually driven near their resonance frequencies where they provide the highest acoustic output. Projectors are sound sources.

■ propagation

the process an electromagnetic wave undergoes as it is radiated from the antenna and spreads out across the physical terrain. See also propagation channel.

■ propagation channel

the physical medium electromagnetic wave propagation between the transmit and receive antennas, and includes everything that influences the propagation between the two antennas.

■ propagation delay

time required for a signal to pass through a device or circuit.

■ propagation time

time required for a wave to travel between two points.

■ proportional amplifier

a particular type of OPERATIONAL AMPLIFIER where the output voltage is in proportion to the difference between it's inputs. Unlike the COMPARATOR, which can based on exactly the same INTEGRATED CIRCUIT, the Op-Amp has two individual inputs instead of one input and one reference value. This use is sometimes called a DIFFERENTIAL AMPLI-

FIER or SUBTRACTOR.

protoboard
board with provision for attatching components without solder. Also called a breadboard. Primarily used for constructing experimental circuits.

proton
sub atomic particle within the nucleus of an atom. Has a positive charge.

proximitor
see Oscillator/Demodulator.

proximity sensor
Ultrasonic sensor designed to measure the distance from the sensor face to a target.

p-type
the conductivity type of a semiconductor material when the majority of carriers are holes, hence positive. P-type silicon is doped with dopant atoms from column III of the periodic table, such as boron.

pulsating current
a unidirectional current that occurs at periodic intervals.

pulse
rise and fall of some quantity (usually voltage) for a period of time.

pulse energy
the power of a single, brief emission from a laser programmed for pulsed behavior rather than continuous operation. Pulse energy can be several times greater than CW emission.

pulse fall time
time for a pulse to decrease from 90% of its peak value to 10% of its peak value.

pulse repetition frequency
the number of times per second that a pulse is transmitted. Pulse rate.

pulse repetition time
time interval between the start of two consecutive pulses.

pulse rise time
time required for a pulse to increase from 10% of its peak value to 90% of its peak value.

pulse tail
pulse decay time, which can be shortened (by using a special mixture of gases) to allow for fast repetition of laser pulses within a given length of time.

pulse width
time interval between the leading edge and trailing edge of a pulse at a point where the amplitude is 50% of the peak value.

pulse width modulation (PWM)

changing the width of an output pulse proportional to an input signal.

pulsed c-v

capacitance vs. Voltage measurements made by sweeping through a series of voltages from start to stop with the voltage returned after each point to a bias voltage level. The bias voltage is usually set in the accumulation or depletion region. Pulsed C-V is used to measure doping profiles deeper into the semiconductor device than with a staircase waveform.

push to make switch

a push to make switch contains two contacts that are bridged or forced together when the switch is operated. As soon as the finger is removed from the switch the contacts are broken. It is an input device .For the switch to turn on pressure must be applied. It is a physical sensor.

push-pull amplifier

an amplifier which uses two transistors (or electron tubes) whose output signals are combined to provide a larger gain (usually a power gain) than a single transistor (or electron tube) can provide.

push-pull circuit

a dual tube or transistor r-f or audio amplifier system in which the signals appearing at each input at any instant are 180° out of phase. The output signal power delivered is the combination of the signal powers of both tubes.

pyrophoric

a chemical that will ignite spontaneously in air at a temperature less than or equal to 54.4°C (130°F).

Pythagorean Theorem

a theorem in geometry. The square of the hypotenuse of a right triangle equals the sum of the squares of the other two sides. In electronics used for vector analysis of AC circuits.

Q

Quality factor of an inductor or capacitor. It is the ratio of a component's reactance (energy stored) to its effective series resistance (energy dissipated). For a tuned circuit, a figure of merit used in bandwidth calculations. Q is the ratio of reactive power to resistive power in a tuned circuit. Also the symbol for charge in coulombs (Q for quantity).

■ Q-switch

a device that has the effect of a shutter moving rapidly in and out of the beam to "spoil" the resonator's normal Q, keeping it low to prevent lasing action until a high level of energy is stored. Result a giant pulse of power when normal Q is restored.

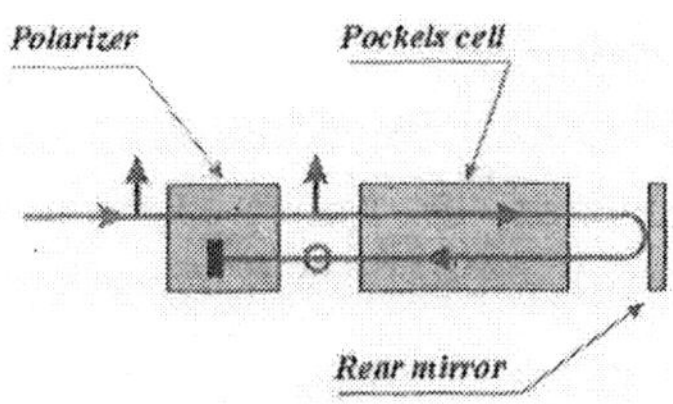

Setup of Q-switch

■ quadrature circuit

a circuit designed to provide a 900 signal phase displacement.

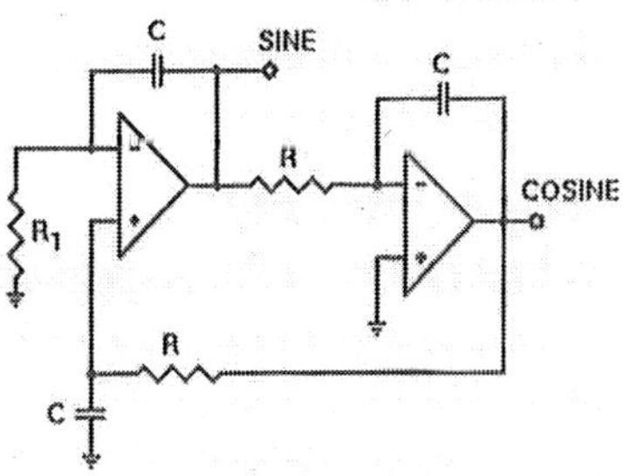

A Quadrature Amplifier

■ quantisation error

when a continuous time signal is digitized, because there isn't an infinite number of discrete digital levels, the difference between the actual analog value and the digital representation of that value is defined as the quantisation error.

■ quantisation

a process where the continuous range of values of an input signal is divided into non-overlapping sub-ranges and, to each sub-range, a discrete value of the output is uniquely assigned.

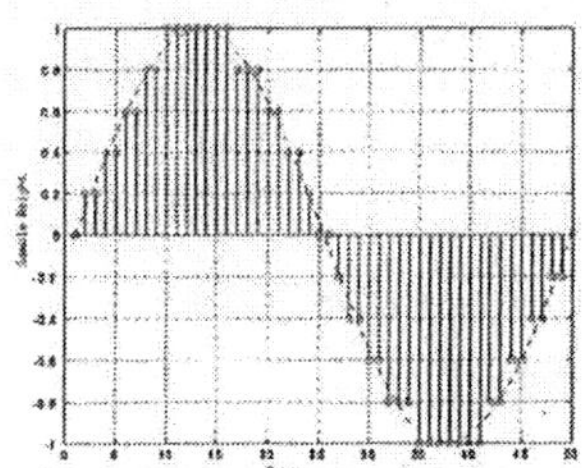

4 it Quantization

■ quasistatic c-v

a Capacitance vs. Voltage measurement obtained using such a low frequency test signal that both minority carriers in the inversion layer and interface traps completely follow the AC gate voltage. Thus, they do contribute to the quasistatic capacitance.

■ **quiescent**
(at rest) for an amplifier the term is used to describe a condition with no active input signal.

■ **quiescent current**
the standing current that flows in a circuit when the signal is not applied. The quiescent current is usually very low or lower than when processing a signal.

■ **quiescent operating point**
the zero-signal voltage and current values in a circuit or amplifier.

■ **quiescent point**
(Q point) a point on the DC load line of a given amplifier that represents the quiescent (no signal) value of output voltage and current for the circuit.

■ **R-2R ladder**
network or circuit composed of a sequence of L networks connected in tandem. Circuit used in digital to analog converters.

■ **RADAR**
acronym for "radio detection and ranging" a system that measures the distance and direction of objects.

■ **radial**
direction perpendicular to the shaft centreline.

■ **radial position**
the average location, relative to the radial bearing centreline, of the shaft dynamic motion.

■ **radial vibration**
shaft dynamic motion or casing vibration which is in a direction perpendicular to the shaft centreline.

■ **radian**
an arc in a circle, equal in length to the radius; an angle (57.3°) at the centre of a circle, formed by 2 radii cutting off such an arc. Thus one rad = 57.3°.

■ **radiance**
brightness; the radiant energy per unit solid angle and per unit projected area of a radiating surface.

■ **radiant energy**
energy traveling as wave motion; specifically, the energy of electromagnetic waves (light, X-rays, radio, gamma rays).

■ **radiant flux**
the rate of emission or transmission of radiant energy.

■ **radiant intensity**
radiant power, or flux, expressed as emission per unit solid angle about the direction of the light in a given length of time.

radiant power
the amount of radiant energy available per unit; the radiant flux.

radiation hardness
the degree to which the characteristics of a device are stable despite exposure of the device to ionising radiation.

radio broadcast
transmission of music, voice and other information on radio carrier waves that can be received by the general public.

radio frequency
RF for short. That part of the spectrum from approx 50kHz to gigahertz.

radio astronomy
branch of astronomy that studies the radio waves generated by celestial bodies and uses these emissions to obtain information about them.

radio communication
term used to describe the transfer of information between two or more points by use of radio or electromagnetic waves.

radio-frequency amplifier
amplifier having one or more active devices to amplify radio signals.

radio-frequency choke
an inductor designed to have a high reactance when used in series with a signal-carrying lead.

radio-frequency generator
Generator capable of supplying RF energy at any desired frequency in the radio-frequency spectrum.

radio-frequency probe
probe used in conjunction with an AC meter to measure radio-frequency signals.

RAM
Random Access Memory; memory arrays which permit reading or writing data in any memory cell selected using a cell address code.

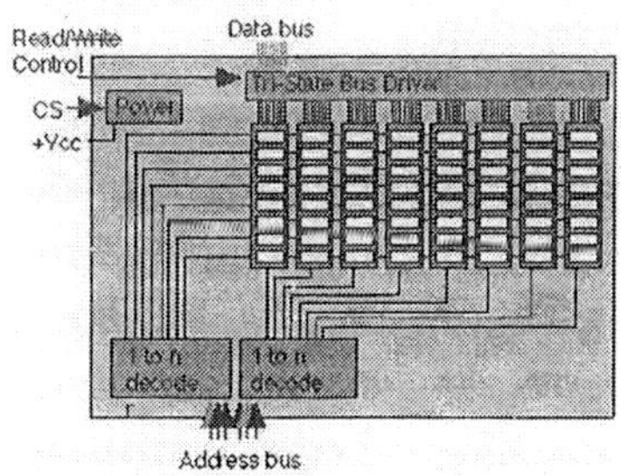

Raman effect
part of the energy in a photon is transferred to (or from) the vibrational/rotational energy of a molecule.

ramp quasistatic c-v

method

the traditional method for measuring quasistatic C-V. In this method, the substrate of an MIS capacitor is swept with an analog ramp and the capacitor's displacement current (I) is measured at the gate. Then, the capacitance is calculated using:
quasistatic C = I (dV/dt).

■ **ramper**

power controls for accomplishing the ramping steps; may be computer-controlled.

■ **range**

a continuous band of signal values that can be measured or sourced. In bipolar instruments, range includes positive and negative values.

■ **rapid thermal processing**

processing semiconductor devices using higher temperatures and shorter processing times. Examples include rapid thermal annealing and rapid thermal oxidation.

■ **rated accuracy**

the limit that errors will not exceed when the instrument is used under specified operating conditions. It is expressed as a percentage (of input or output) plus a number of counts.

■ **ratio detector**

a dual-diode frequency-modulation detector.

■ **ratio measurement**

the measurement of a signal input with relation to an external reference input.

■ **ratiometric measurement (referring to an ADC)**

rather than using a voltage reference with an absolute value, a ratio of the signal applied to the transducer (that is, the load cell or bridge) is also applied to the voltage reference input of the analog-to-digital converter (ADC), thereby eliminating any errors introduced by a changing reference.

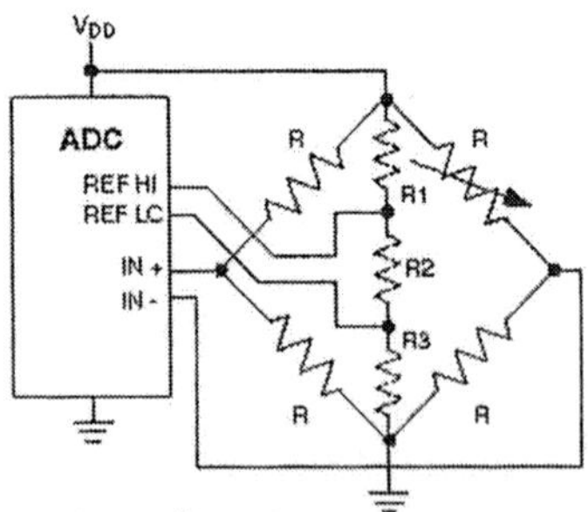

■ **rationalisation**

the process of trimming the sensor output at the rated input to a target value.

R-C oscillator
a circuit whose output is an AC signal (square wave or sine wave) at a fixed frequency (oscillation) set by an external resistor and capacitor. Changing R or C changes the output frequency. Most general purpose timers are R-C Oscillators.

RC time constant
Product of resistance and capacitance in seconds.

reactance
opposition to alternating current flow (radio energy) by storage in an electrical field (capacitive reactance) or in a magnetic field (inductive reactance). Measured in ohms (Ω). Symbol capacitive -X, inductive +X.

reactance circuit
a circuit designed to behave as either an inductive or capacitive reactance, and hence, as an equivalent inductor or capacitor.

reactive power
also called imaginary power or wattless power. It is the power value in "volt amps" obtained from the product of source voltage and source current in a reactive circuit.

reading rate
the rate at which the displayed number is updated.

reading
the displayed number that is proportional to the measured magnitude of the input signal.

real number
number having no imaginary part.

real-time analyser
see Dynamic Signal Analyser.

real-time rate
For a DSA, the broadest frequency span at which data is sampled continuously. Real-time rate is mostly dependent on FFT processing speed. If the definition of real time rate is "not miss any data", the real-time rate will be window dependent. The real-time rate will decrease when using any other window than uniform.

receive cable
a known good fiber optic jumper cable attached to a power meter used as a reference cable for loss testing. This cable must be made of fiber and connectors of a matching type to the cables to be tested.

receiver
transducer used to intercept the acoustic wave reflected back from the target. Can be same as transmitter.

receiver (Rx)
a circuit in which electrical waves or currents are converted into audible signals.

reciprocity
reciprocal behaviour of a system. The conditions for matching a transmitter to the antenna are the same as those matching the antenna to a transmitter. a well matched system will exhibit better receive characteristics than a mismatched one.

recombination
process by which a conduction band electron gives up energy (in the form of heat or light) and falls into a valence band hole.

recovery time
the time required for an MIS capacitor to reach equilibrium in inversion after being pulsed from accumulation or depletion into deep depletion. The recovery time is directly proportional to the generation lifetime of the MIS device.

rectangular coordinates
a Cartesian coordinate of a Cartesian coordinate system whose straight-line axes or coordinate planes are perpendicular.

rectangular wave
also known as a pulse wave. a repeating wave that only operates between two levels or values and remains at one of these values for a small amount of time relative to the other value.

rectangular window
see Uniform Window.

rectification
a function that changes a bipolar signal to a unipolar signal. Often the function is implemented with a reference and a comparator.

rectifier
a device that pases current in only the forward direction.

reed relay
relay consisting of two thin magnetic strips within a glass envelope. When a coil around the envelope is energised, the relay's contacts snap together making a connection between leads attached to the reed strips.

reed switches
a reed switch, sometimes called a proximity sensor reacts when a magnet is moved close to it. Inside the case there is a thin piece

of metal that moves, touching another piece of metal. This makes the contact for the current to flow.It is an input device .For the switch to turn on a magnet must beplaced near the switch. It is a remote device.

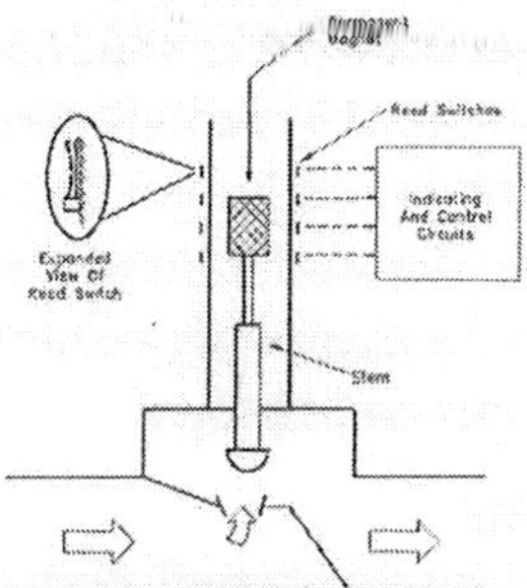

Construction of Reed Switch

■ **reference mark**

random graduation pattern which - when traversed over - produces a signal peak, which may be used to determine an absolute datum within an incremental measuring system.

■ **reference pulse**

square-wave signal produced when the scale reference mark is traversed over; normally one measuring step wide; may be used to define an absolute datum within an incremental measuring system.

■ **reflectance**

the ratio of the reflected flux to the incident flux, or the ratio of reflected light to light falling on the object.

■ **reflection**

a process that occurs when a propagating electromagnetic wave impinges upon a obstruction whose dimensions are very large when compared to the wavelength. Reflections from the surface of the earth, and from buildings or walls produce reflected waves which may interfere, constructively of destructively at the receiver.

■ **refraction**

the change of direction of propagation of any wave, such as an electromagnetic wave, when it passes from one medium to another in which the wave velocity is different. Simply put, the bending of incident rays as they pass from one medium to another, such as air to water.

■ **refractive index**

a property of optical materials that relates to the velocity of light in the material.

■ **regeneration**

the process for providing an internal positive feedback within an electronic system for increasing signal amplitudes. Excessive regeneration produces oscillation and, hence, signal generation.

■ **regenerative feedback**

positive feedback. Feedback from the output of an amplifier to the input such that the feedback signal is in phase with the input signal. Used to produce oscillation.

■ **regenerative receiver**

an amplitude modulated (AM) radio receiver in which positive feedback is used in order to increase the sensitivity and selectivity of the receiver.

■ **regulated power supply**

power supply that maintains a constant output voltage under changing load conditions.

■ **regulation**

the degree by which a voltage source is held near its no-load amplitude when a load is applied.

■ **regulator**

device or circuit that maintains a desired output under changing conditions.

■ **relative**

not independent. Compared with or with respect to some other measured quantity.

■ **relative accuracy**

see integral non-linearity (INL).

■ **relative motion**

Vibration measured relative to a chosen reference. Displacement transducers generally measure shaft motion relative to the transducer mounting.

■ **relaxation oscillator**

an oscillator circuit which generates a signal by periodic conduction-non-conduction periods. a non-sinusoidal waveform is produced and, generally, no resonant circuits are employed.

■ **relay**

1. electromechanical device that opens or closes contacts when a current is passed through a coil.
2. these are very useful but expensive electrical switches that can control very large amounts of electrical energy. a relay consists of a switch that is operated by an electrical magnet. There are several types, including normally-open (NO) and normally-closed (NC). In an NO relay, when the magnet is on the metal parts of the switch are pulled together. When the magnet is off the metal parts of the switch are pushed apart. The action is the opposite way around for an NC relay. This is an example of a switch containing moving parts.

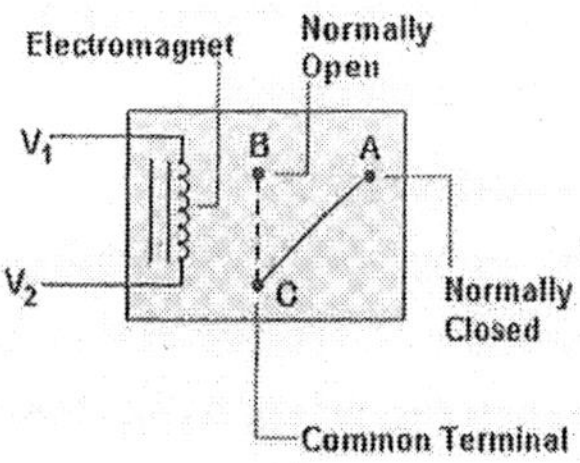

SPDT Relay

■ **relay must-operate value**
a specified functioning value at which all relays meeting the specifications must operate. Also known as relay pick-up or pull-in value.

■ **relay must-release value**
a specified functioning value at which all relays meeting the specifications must release. Also known as relay drop-out value.

■ **release time**
the time between the removal of the relay coil voltage and the stabilized opening of the relay contacts.

■ **reliability**
the ability of a device to perform within the desired range over a measured period of time.

■ **reluctance**
resistance to the flow of magnetic lines of force.

■ **reluctance**
the opposition offered by a material to magnetic flux.

■ **reluctance**
the opposition offered by a material to magnetic flux.

■ **remanence**
amount a material remains magnetized after the magnetizing force has been removed.

■ **remote cutoff tube**
sometimes known as a variable-mu tube. It is characterized by a wide negative-bias latitude during which current flows.

■ **remote input coupler**
the Keithley Model 5951 Remote Input Coupler is a device that connects the 595 Quasistatic C-V Meter and the 590/100k/1M C-V Analyser together to perform simultaneous C-V. The coupler uses precision RLC circuits to combine the test signals from and route the displacement current to the two meters.

■ **repeatability**
the ability of a transducer or readout instrument to reproduce readings when the same input is applied repeatedly.

■ **repeater, regenerator**
a device that receives a fiber optic signal and regenerates it for retransmission, used in very long

fiber optic links.

residual magnetism

the flux density remaining in a material after the magnetizing force has been removed.

resistance

the measure of the ability of a material to pass a current.

resistance

opposition to current flow by conversion into other forms of energy, such as heat. Measured in ohms (Ω). Symbol R.

resistance insertion

a current measuring technique where a known resistor is connected in series with the circuit to be measured. The voltage drop across the resistor is proportional to the unknown current.

resistive power

amount of power dissipated as heat in a circuit containing resistive and reactive components. True power as opposed to reactive power.

resistive temperature detector

(RTD) Temperature detector consisting of a fine coil of conducting wire (such as platinum) that will produce a relatively linear increase in resistance as temperature increases.

resistivity

the geometry independent parameter of a material that determines the relationship between the current through it and the current across it. Resistance of a cylinder of material is: resistance = resistivity * (length / area).

resistor (fixed)

resistors are used to restrict the amount of CURRENT flowing through a circuit. They are often used to protect other delicate devices, such as LEDs, from damaging high voltages. The value of a resistor is measured in OHMS and we can use this to calculate how much current will pass through it. High value resistors may become very hot and these may be fitted with fins to help stop them melting. There are two special types, THERMISTORS and LDRs (sometimes called PHOTOCONDUCTORS). a single fixed resistor restricts the flow of current in a circuit. It is a passive component .Resistance is measured in Ohms .Circuit symbol for a resistor.

resistor colour code

coding system of colour stripes

on a resistor to indicate the resistor's value and tolerance.

■ **resistor**
passive component with a known resistance. The value of resistance is usually shown by a set of coloured bands on the body of the component.

■ **resolution**
when an analog signal is digitized, it is represented by a finite number of discrete voltage levels. The resolution is the number of discrete levels that are used to represent the signal. To more accurately replicate the analog signal, the resolution must be increased. Resolution is usually defined in bits. Using converters with higher resolutions will reduce the quantisation error. *See also* quantisation error.

■ **resolution**
the smallest change in stimulus that will produce detectable change in the instrument output.

■ **resolution (data acquisition)**
the smallest signal increment that can be detected by a measurement system. It is usually specified in "bits."

■ **resolution**
the smallest value of input (or output) signal, other than zero, that can be measured (or sourced) and displayed. Also called sensitivity or minimum resolvable quantity.

■ **rocolver**
inductive angle measuring device, producing two alternating voltages whose amplitudes or phases depend on the (shaft) rotation angle.

■ **resonance**
the condition of vibration amplitude and phasechange response caused by a corresponding system sensitivityto a particular forcing frequency. a resonance is typically identifiedby a substantial amplitude increase, and related phase shift.

■ **resonant frequency**
the frequency at which a piezoelectric ceramic will vibrate most efficiently i.e. will produce the highest output with the least amount of voltage applied.

■ **resonator**
the mirrors (or reflectors) making up the laser cavity containing the laser rod or tube. The mirrors reflect light back and forth to build up amplification under

an external stimulus. Emission is through one of them, called a coupler, which is partially transmissive.

■ **response time**

for a measuring instrument, the time between application of a step input signal and the indication of its magnitude within a rated accuracy. For a sourcing instrument, the time between a programmed change and the availability of the value at its output terminals. Also known as Settling Time.

■ **retentivity**

the degree by which a material retains magnetism.

■ **reversal error**

measuring error which results from approaching a position from different directions.

■ **reverse battery protection**

a circuit designed to protect an ASIC from being destroyed when supplied with the wrong polarity.

■ **reverse bias**

bias on a PN junction that allows only leakage current (minority carriers) to flow. Positive polarity on the n-type material and negative polarity to the p-type material.

■ **reverse breakdown voltage**

amount of reverse bias that will cause a PN junction to break down and conduct in the reverse direction.

■ **reverse current**

current through a diode when reverse biased. an extremely small current also referred to as leakage.

■ **reverse engineering**

the process of determining the actual circuit by study of the geometrical layout of a fabricated chip.

■ **reverse saturation current**

reverse current through a diode caused by thermal activity. This current is not affected by the amount of reverse bias on the component, but does vary with temperature.

■ **RF**

radio Frequency; electrical voltages which oscillate at frequencies and in a manner which make possible their electromagnetic transmission through space.

■ **RF (radio frequency)**

amplifier

an amplifier designed to amplify signals with frequencies between 10 kilohertz (10 kHz) and 100,000 megahertz (100,000 MHz).

RF (radio frequency) transformer

a transformer specially designed for use with rf (radio frequencies). an rf transformer is wound onto a tube of non- magnetic material and has a core of either powdered iron or air.

RF continuity

the ability of a system to provide a closed circuit for the RF current to provide RF radiation.

rheostat

a variable resistor having two terminals, as distinguished from the potentiometer where three terminals are used.

right angle triangle

triangle having a 90° or square corner.

ringing

analogous to the ringing of a bell, it is the rise and decay time before and after the transducer reaches maximum amplitude. Expressed as the mechanical Q of the transducer which is the number of cycles it takes to get up to 90% of maximum amplitude, or down to 10% above zero amplitude.

ripple frequency

Frequency of the ripple present in the output of a DC source.

ripple voltage

the small variations in Dc voltage that remain after filtering in a power supply.

rise time

the time required for a signal to change from a small percentage (usually 10%) to a large percentage (usually 90%) of its peak-to-peak amplitude.

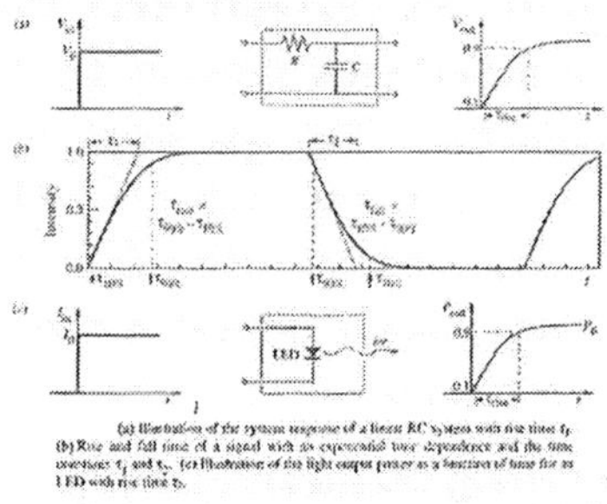

Comparision Between Rise Time and Fall Time

RL differentiator

an RL circuit whose output voltage is proportional to the rate of change of the input voltage.

RL filter

selective circuit of resistors and inductors that offers little or no opposition to certain frequencies while blocking or attenuating other frequencies.

■ RL integrator

RL circuit with an output proportionate to the integral of the input signal.

■ rms responding

a measurement where the displayed value is equal to the root-mean-square (rms) of the input signal, for all input waveforms having components within the specified frequency range and crest factor limit.

■ rms value

rms value of an AC sine wave is 0.707 times the peak value. This is the effective value of an AC sine wave. The rms value of a sine wave is the value of a DC voltage that would produce the same amount of heat in a heating element.

■ rolling element bearing

Bearing whose low friction qualitiesderive from rolling elements (balls or rollers), with little lubrication.

■ roll-off rate

rate of change in gain when an amplifier is operated outside of its bandwidth.

■ ROM

Read Only Memory; an electronic or disk-based data storage medium in which preprogrammed data can be accessed as required by a system.

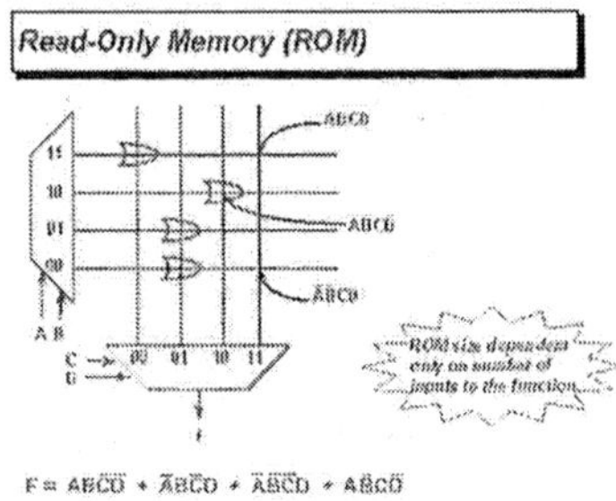

■ root mean square (rms)

the effective value or effective DC value that an AC signal represents. For a sine wave, the RMS value is 0.707 times the peak value, or 0.354 times the peak-to-peak value.

■ rotary switch

electromechanical device that has a rotating shaft connected to one terminal capable of making or breaking a connection to one or more other terminals.

■ rotating lens

beam delivery in a circular movement for cutting large-diameter holes. (De-focusing the beam for this lowers power density and increases drilling time.).

■ **rotor, flexible**

a rotor which operates close enough to, or beyond its first bending critical speed for dynamic effects to influence rotor deformations. Rotors which cannot be classified as rigid rotors are considered to be flexible rotors.

■ **rotor, rigid**

a rotor which operates substantially below its first bending critical speed. a rigid rotor can be brought into, and will remain in, a state of satisfactory balance at all operating speeds when balanced on any two arbitrarily selected correction planes.

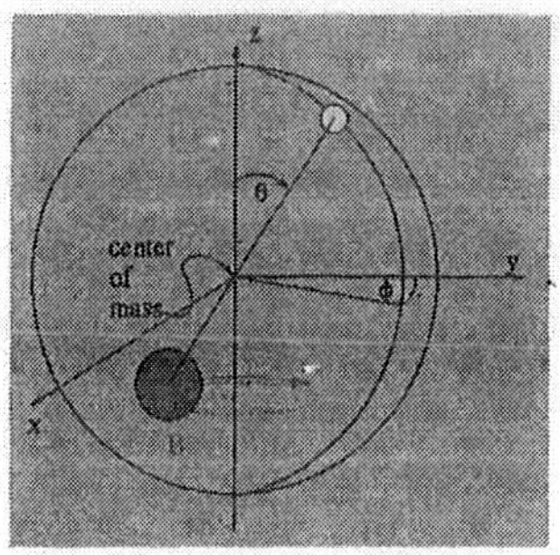

■ **runout compensation**

electronic correction of a transducer output signal for the error resulting from slow roll run out.

■ **sample and hold amplifier**

an op amp whose input may be turned on to sample a voltage, then turned off to hold the sampled voltage at the output, while an ADC converts the voltage to a digital number.

■ **sample draw method (gas monitoring)**

air samples are drawn from the point of measurement into the detector head across the sensor, using a diaphragm or eductor pump.

■ **sample rate**

the speed specification of an A/D or D/A converter. It describes the maximum data throughput and is measured in samples/second or Hertz. an example would be 100ks/s, or 100kHz.

■ **sample**

a value that is read from or written to one channel.

■ **sampling rate/sampling frequency**

the rate at which a converter acquires the input signal, digitises it, and outputs data to a ƒÊC/DSP. It is specified in samples per second or Hertz (Hz) and is also referred to as the "throughput rate."

■ **saturable-core reactor**

a coil whose reactance is controlled by changing the permeability of the core.

saturation
condition in which a further increase in one variable produces no further increase in the resultant effect. In a bipolar junction transistor, the condition when the emitter to collector voltage is less than the emitter to base voltage. This condition puts forward bias on the base to collector junction.

saturation
in a tube or transistor, the point at which current levels off for additional increases in voltage. For magnetism, the point at which flux density levels off for additional increases in magnetizing force.

saturation (magnetic core)
the condition in which a magnetic material has reached a maximum flux density and the permeability has decreased to a value of (approximately) 1.

saturation currents. (parametric testing)
a CMOS device parameter. Drain current at maximum drain and gate voltages.

sawtooth voltage
a waveform characterized by a gradual incline and an abrupt decline of amplitude, so that successive signals produce a sawtooth pattern.

sawtooth wave
repeating waveform that rises from zero to maximum value linearly drops back to zero and repeats. a ramp waveform.

scale
set of markings used for measurement.

scan (data acquisition)
sample a group of channels once at the acquisition or burst-mode rate; also can refer to a group of channels. These channels may be sequential (start to stop channel specified) or nonsequential (channel-gain queue used).

scan rate
the rate at which a group of channels is sampled, measured from the start of one scan to the start of the next.

scan
sequential connecting (usually break-before-make) of one instrument to multiple devices under test or multiple instruments to one device under test..

scanning frequency
response level which limits the velocity of an incremental measuring system.

■ **scattering**
the change of direction of light after striking small particles that causes loss in optical fibers.

■ **schematic diagram**
illustration of an electrical or electronic circuit with the components represented by their symbols.

■ **Schmitt trigger**
a building block in which the input must be taken to about 70% of rail voltage before the output will change. The lower level for change is about 30%. This produces a gap and thus noise is prevented from entering.

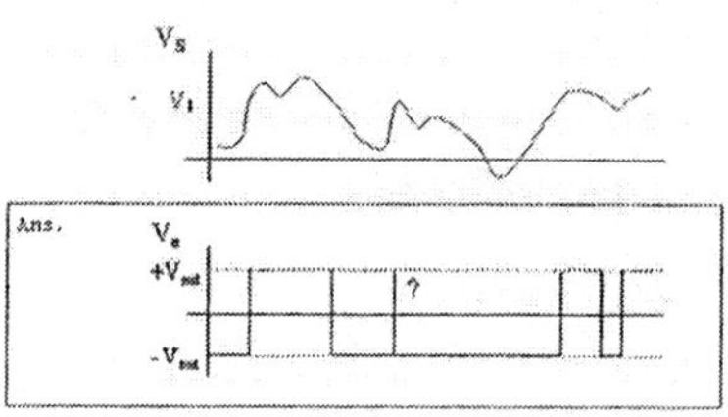

Schmitt Trigger Circuit Response

■ **Schottky diode**
a junction or barrier formed by the direct contact of semiconductor materials with a metal. This type of contact rectifies signals and may also exhibit some resistance.

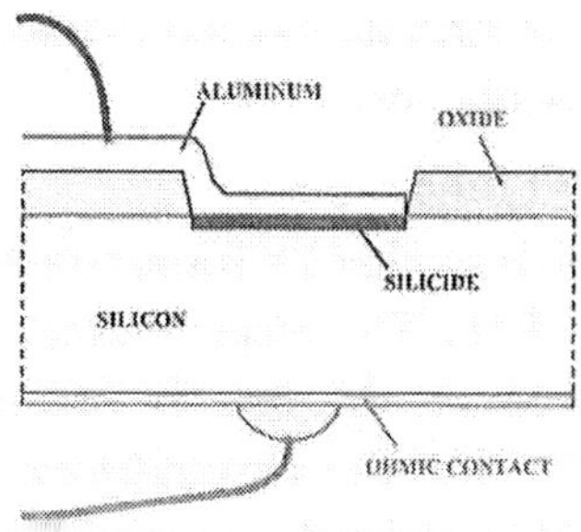

Schematic of a Silicide Schottky Diode Fabricated on Silicon

■ **scientific notation**
numbers entered as a number from one to ten multiplied by a power of ten. Example: 8765 = 8.765 × 103.

■ **screen grid**
the electron-beam accelerating electrode in a tube. This grid is usually placed at signal ground to act as an electrostatic shield between input and output sections.

■ **scribe and break**
the procedure used to separate a processed wafer into individual ICs. Narrow channels between individual ICs are mechanically weakened by scratching with a diamond tip (scribe), sawing with a diamond blade, or burning with a laser. The wafer is mechanically stressed and broken apart along the channels (called scribe lines), thereby separating the individual ICs (dice). Test

structures are sometimes placed in scribe lines.

■ SECAM.

this is another TV format similar to PAL. The major difference between the two is that in SECAM, the chrominance is FM-modulated.

■ secondary

the output side of a transformer. a re-chargeable battery such as a car battery. Also called accumulator. The battery does not make electricity but simply stores energy in chemical form.

■ secondary cell

electrolytic cell used to store electricity. Once discharged may be restored by recharging by putting current through the cell in the direction opposite to that of discharge current.

■ secondary emission

electron emission from a non heated electrode when struck by a high-velocity electron beam.

■ seismic

refers to an inertially referenced measurement or a measurement relative to free space.

■ seismic transducer

a transducer that is mounted on the case or housing of a machine and measures casing vibration relative to free space. Accelerometers and velocity transducers are seismic.

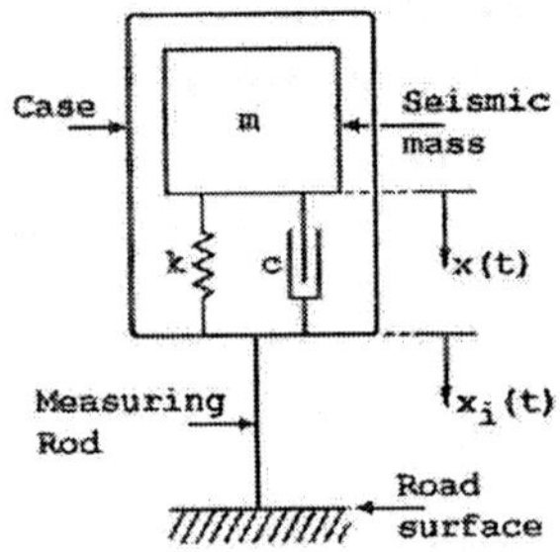

Seismic Transducer Working

■ selectivity

characteristic of a circuit to discriminate between wanted and unwanted signals.

■ self- bias

bias developed from the supply rail due to circuit components and a dropper resistor rather than a separate battery.

■ self biasing

gate bias for a field effect transistor in which source current through a resistor produces the voltage for gate to source bias.

■ self inductance

property that causes a counterelectromotive force to be produced in a conductor when the magnetic field expands or collapses with a change of current.

■ **semiconductor**

an element which is neither a good conductor or a good insulator, but rather lies somewhere between the two. Characterized by a valence shell containing four electrons. Silicon, germanium and carbon are the semiconductors most frequently used in electronics.

■ **semiconductor**

a solid state material that can be physically altered to modify its electronic properties. a semiconductor is half way between a conductor and an insulator. When charged with electricity or light, semiconductors change their state from non-conductive to conductive or vice versa.

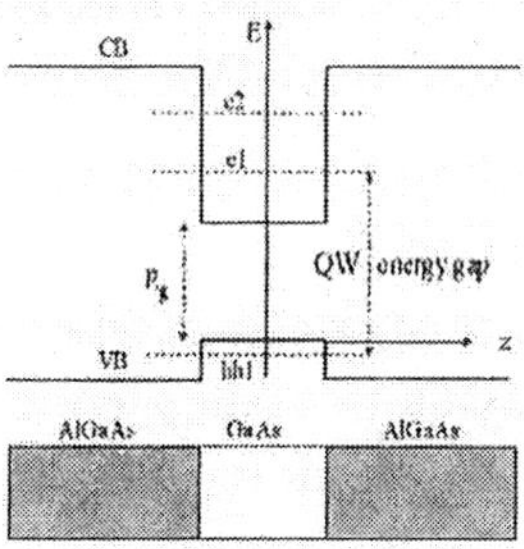

A Quantum Well Structure

■ **semiconductor**

a class of materials (e. g. silicon, gallium arsenide) that exhibit electrical properties mid-way between insulators (e.g., plastic) and conductors (e. g., copper).

■ **semiconductor band gap**

the amount of energy needed to move an electron in a semiconductor material from the top of the valence band to the bottom of the conduction band. In silicon, this is about 1. 12V.

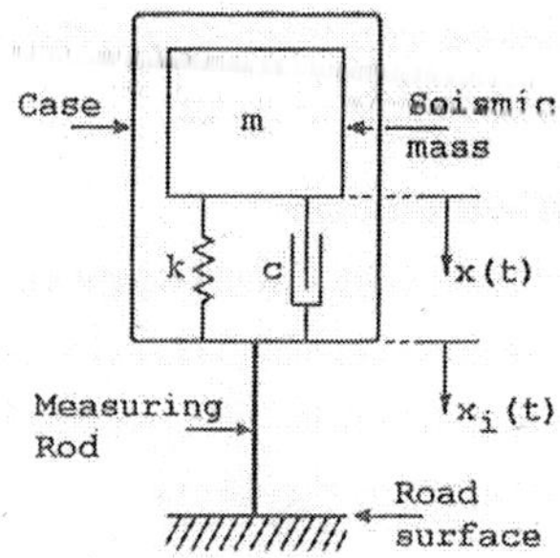

■ **semiconductor**

an element such as silicon or germanium or a compound like GaAs that has an intermediate band gap. Unlike metals that freely conduct and insulators that do not conduct charge, semiconductors selectively conduct charge through the movement of holes and electrons.

■ **sensitivity**

a measure of the ability of a receiver to respond to weak signals.

■ **sensor**

the combination of a transducer and some signal-conditioning circuitry. The transducer converts some physical condition such as temperature, pressure, humidity, etc. into an electrical signal (typically a voltage or current). The signal-conditioning circuitry might then amplify and filter, or do other types of corrections.

■ **series**
resistors, cells, switches and batteries etc can be connected in series as in this diagram

■ **series circuit**
circuit in which the components are connected end to end so that current has only one path to follow through the circuit.

■ **series parallel network**
network that contains components connected in both series and parallel.

■ **series peaking**
a technique used to improve high-frequency response in which a peaking coil is placed in series with the output signal path.

■ **series RC model**
interpretation of the real and imaginary components of admittance measured by a capacitance meter as resistance in series with capacitance.

■ **series resistance**
the resistance between the substrate contact to an MIS device and the MIS capacitor. This resistance is increased when substrate doping is low or backside contact is poor.

■ **series resonance**
condition that occurs in a series LC circuit at the frequency where inductive reactance equals capacitive reactance. Impedance is minmum, current is maximum limited only by resistance in the circuit.

■ **series resonant circuit**
a circuit in which the components are connected in series and capable of oscillating at a particular frequency. It exhibits minimum impedance at resonance.

■ **settling time (data acquisition)**
time it takes for a voltage to settle and remain within a specified error band around the final value.

■ **settling time (referring to the DAC)**
the actual amount of time it takes a digital-to-analog converter (DAC) output to reach its final value (within a certain percentage) once the DAC has accepted a command to change its output value. This time can be affected by the slew rate of an output amplifier and by the amount of amplifier ringing and signal overshoot.

■ **settling time**

For a measuring instrument, the time between application of a step input signal and the indication of its magnitude within a rated accuracy. For a sourcing instrument, the time between a programmed change and the availability of the value at its output terminals. For a switching card, the time required for establishing relay connections and stabilizing user circuits. Also known as Response Time.

■ **seven segment display**

device made of several light emitting diodes arranged in a numeric or alphanumeric pattern. By lighting selected segments numeric or alphabet characters can be displayed.

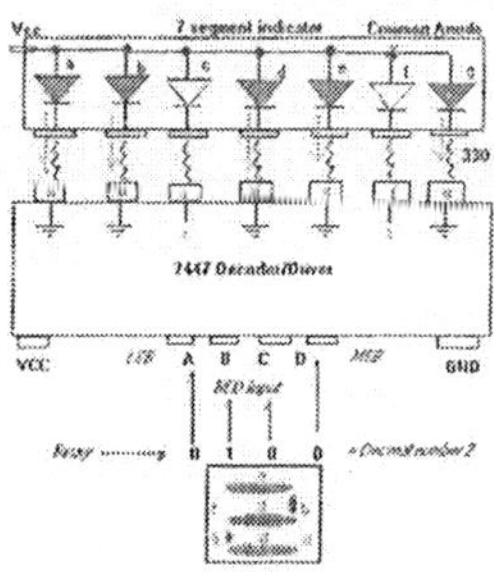

Construction of 7 Segment display using A 7447 Decoder

■ **sheet resistance**

a measure of the resistance of a film of material related to its resistivity divided by the film thickness. It is often measured using a four-point probe and is directly related to the average doping in the film.

■ **sheet resistances. (parametric testing)**

a CMOS or bipolar device parameter. Resistance per square of conductor.

■ **shells or bands**

orbital path containing a group of electrons having a common energy level.

■ **shield**

metal grounded cover used to protect a wire, component or piece of equipment from stray magnetic and/or electric fields.

■ **shielding**

a metal enclosure for the circuit being measured or a metal sleeve surrounding wire conductors (coax or triax cable) to lessen interference, interaction, or current leakage. The shield is usually grounded.

■ **short circuit**

also called a short. Low resistance conection between two points in a circuit typically causing excessive current.

■ **short wavelength**

a commonly used term for light in the 665, 790, and 850 nm ranges.

■ **short-circuit**

an un-intended path that conducts electricity. Also called a "bridge" or "short" such as when solder from two tracks touch on a PC board.

■ **short-term accuracy**

the limit that errors will not exceed during a 24-hour period of continuous operation. Unless specified, no zeroing or adjustments of any kind are permitted. It is expressed as a percentage of reading (or sourced value) plus a number of counts over a specified temperature range.

■ **shunt**

this generally means parallel connection.

■ **shunt capacitance loading**

the effect on a capacitance meter of capacitance, such as from cables or fixtures, between the meter input or output terminals and guard (sometimes referred to as common). Shunt capacitance reduces the measured device current. Therefore, the measured device capacitance is smaller.

■ **shunt peaking**

a technique used to improve high-frequency response in which a peaking coil is placed in parallel (shunt) with the output signal path.

■ **shunt resistor**

resistor connected in parallel or in shunt with another component or circuit.

■ **side lobe**

smaller acoustic beams located around the main lobe.

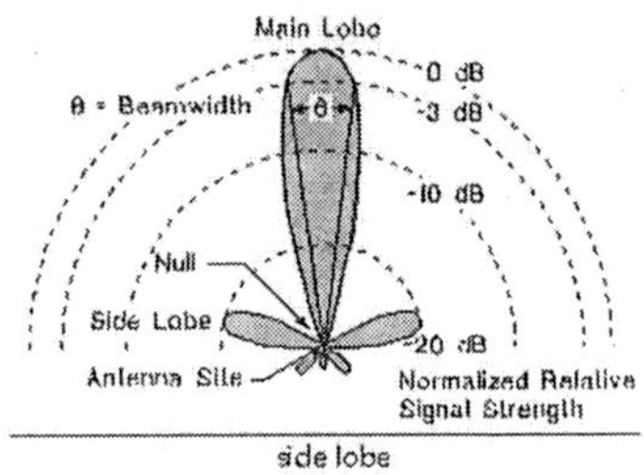

Antenna's Side Lobe

■ **signal**

a general term used to describe any a.c. or d.c. of interest in a circuit; e.g., input signal.

■ **signal conditioner**

a device placed between a signal sourceand a readout instrument to change the signal and/or bandwidth.

Example sattenuators, preamplifiers, charge amplifiers, filters.

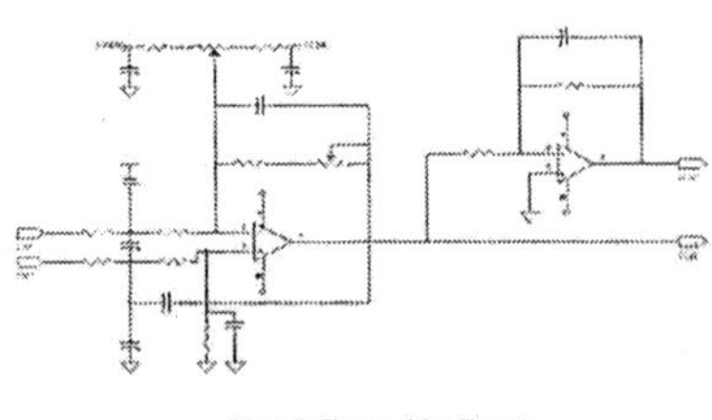

Analog Signal Conditioner

signal generator
a circuit that produces a variable and controllable signal.

signal/noise ratio
the ratio of the maximum signal that can be measured to the level detected with no signal present (noise level). It is expressed in decibels.

signal-to-noise and distortion ratio (SINAD)
the RMS value of the sine wave f(IN) (input sine wave for an ADC, reconstructed output sine wave for a DAC) to the RMS value of the noise of the converter from DC to the Nyquist frequency, including harmonic content. It is typically expressed in decibels. *See also* root mean square (RMS) *and* total harmonic distortion.

signal-to-noise ratio (SNR)
the RMS value of the sine wave f(IN) (input sine wave for an ADC, reconstructed output sine wave for a DAC) to the RMS value of the noise of the converter from DC to Nyquist frequency, excluding noise at DC and harmonic distortion content. It is typically expressed in decibels. *See also* root mean square (RMS).

signature
term usually applied to the vibration frequency spectrum which is distinctive and special to a machine or component, system or subsystem at a specific point in time, under specific machine operating conditions, etc. Used for historical comparison of mechanical condition over the operating life of the machine.

signed binary coding
a coding scheme in which the MSB represents the sign (positive or negative) of a binary number. In this scheme, "-2" is represented by 10000010 and "2" is represented by 00000010.

silicon controlled rectifier
a reverse blocking triode thyristor. a pnp device.

silicon controlled switch
an SCR with an added terminal called an anode gate. a positive pulse either at the anode gate or the cathode gate will turn the device on.

silicon dioxide
glass like material used as the gate insulating material in a MOSFET.

silicon transistor
a bipolar junction transistor us-

ing silicon as the semiconducting material.

silicon
the most common elemental semiconductor. From silicon crystals that are grown and sliced into wafers.

silicon-controlled rectifier (SCR)
three terminal active device that acts as a gated diode. The gate terminal is used to turn the device on allowing current to pass from cathode to anode.

silver mica capacitor
Mica capacitor with silver deposited directly onto the mica sheets instead of using conductive metal foil.

silver solder
solder composed of silver, copper and zinc. Has a melting point lower than pure silver, but higher than lead-tin solder.

simplex
communication in only one direction at a time. Example: FAX.

simulation
the use of a CAD tool which mimics the behavior of the actual circuit through a collection of sub-circuit models and so enables designers to test and debug circuits before layout and fabrication.

simulcast
broadcasting a program simultaneously in two different forms, for example a program broadcast in both AM and FM.

simultaneous sample-and-hold
operation in which the analog input channels are sampled at the same time and the values held until sequentially read by a scanning A/D system.

sinad
see signal-to-noise and distortion ratio (SINAD).

sine
sine of an angle of a right angle triangle is equal to the opposite side divided by the hypotenuse.

sine wave
wave whose amplitude is the sine of a linear function of time. It is plotted on a graph that plots amplitude against time or radial degrees relative to the angular rotation of an alternator.

single in-line package
package containing several electronic components (generally resistors) with a single row of connecting pins.

single pole double throw
(SPDT) three terminal switch in which one terminal can be connected to either one of the other terminals.

single pole single throw
(SPST) two terminal switch or relay thet can open or close one circuit.

single pole switch
a switch in which only one circuit is controlled.

single sideband
(SSB) AM radio communication technique in which the transmitter suppresses one sideband and therefore transmits only a single sideband.

single supply op amp
an op amp operating off a single supply voltage (typically +5V or +3V) vs. one requiring dual supplies (e.g., +/-5V or +/-15V).

single throw switch
switch containing only one set of contacts which can be either opened or closed.

single-ended input
an input signal whose reference is a fixed level, usually ground. Noise that is coupled into the signal will directly add to a measurement error.

single-ended input (data acquisition)
an analog input with one input terminal whose value is measured with respect to a common ground.

single-ended
the condition where the low terminal of a two-terminal instrument is connected to a specific reference point, such as power line common, earth ground, or circuit common.

single mode fiber
a fiber with a small core, only a few times the wavelength of light transmitted, that only allows one mode of light to propagate. Commonly used with laser sources for high speed, long distance links.

sink
device such as a load that consumes power or conducts away heat.

sintering
process of bonding either a metal or powder by cold pressing it into a desired shape and then heating to form a strong cohesive body.

sinusoidal
Varying in proportion to the sine

of an angle or time function. AC voltage in which the instantaneous value is equal to the sine of the phase angle times the peak value.

■ SIP

abbreviation for "sihgle in-line package."

■ skin effect

tendancy of high-frequency (rf) currents to flow near the surface layer of a conductor.

■ slave

a mainframe that is externally connected to a controlling mainframe (master). a slave unit adds capacity or functions to the master..

■ slew rate

the maximum rate at which the digital-to-analog converter (DAC) output can change, or the maximum rate at which an input signal can change without resulting in an error in the digitized representation of the input signal.

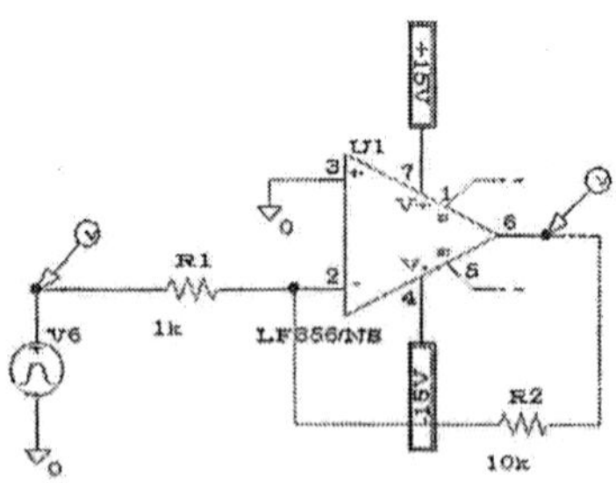

Demonstrates the Finite Slew Rate of an Inverting Amplifier

■ slide switch

a slide switch operates by moving a piece of metal over metal contacts to make the connections. It is an input device It is a physical sensor Robust and suitable for general-purpose applications.

■ slow acting relay

slow operating relay that when energized may not pull up the armature for several seconds.

■ slow roll speed

Low rotative speed at which dynamic motion effects from forces such as imbalance are negligible.

■ slow-blow fuse

Fuse that can withstand a heavy current (up to ten times its rated value) for a small period of time before it opens.

■ small signal level

the amplitude of AC test signal voltage producing a linear response of the device under test.

In other words, over the small signal range, the shape of the device response curve can be assumed to be linear without distorting the results. For MIS devices, this is usually 50mV or less.

■ small-signal bandwidth

a small-amplitude signal is applied to the analog-to-digital converter (ADC) in such a way that the signal's slew rate will not limit the ADC's performance, and the input frequency is swept up to the point where the amplitude of the digitized conversion result has decreased by -3dB. The small-signal bandwidth is often limited by the track/hold amplifier performance.

■ snap switch

switch containing a spring under tension or compression that causes the contacts to come together suddenly when activated.

■ SNR

abbreviation for "signal to noise ratio."

■ soft magnetic material

Ferromagnetic material that is easily demagnetised.

■ software

Computer programs and data which are used in conjunction with hardware to implement a complete microelectronic system (e.g., a personal computer or a telecommunications network).

■ software trigger

a programmed event that starts an operation such as data acquisition.

■ SOI

Silicon-On-Insulator. a composite structure consisting of an active layer of silicon deposited on an insulating material. The insulator can be sapphire, silicon dioxide, silicon nitride, or even an insulating form of silicon itself.

■ solar cell

photovoltaic cell that converts light into electric energy. Especially useful as a power source for space vehicles.

■ solder

metallic alloy used to join two metal surfaces.

■ soldering

process of joining two metallic surfaces to make an electrical contact by melting solder (usually tin and lead) across them.

■ soldering iron

tool with an internal heating element used to heat surfaces being soldered to the point where the solder becomes molten.

■ **solenoid**

an air core coil. Equipped with a movable iron core the solenoid will produce motion. As a result of current through the coil the iron core is pulled into the centre of the winding. When the coil is deenergised, a spring pulls the movable core away from the centre of the winding. Machanical devices connected to the movable core are made to move as a result of current through the coil. Example: Electric door locks on some automobiles.

■ **solid conductor**

Conductor having a single solid wire instead of strands of fine wire twisted together.

■ **solid state**

pertaining to circuits where signals pass through solid semiconductor material such as transistors and diodes as opposed to vacuum tubes where signals pass through a vacuum.

■ **solvent**

a material that is capable of dissolving another chemical to form a uniformly dispersed mixture (solution).

■ **sonar**

word is derived from "sound navigation and ranging." It describes a devise that transmits frequency sound waves in water and registers the vibrations reflected back from an object. It is used in detecting objects such as submarines, locating schools of fish, or determining water depth.

■ **sonic**

pertaining to sound.

■ **sound wave**

pressure waves propagated through air or other plastic media. Sound waves are generally audible to the human ear if the frequency is between approximately 20 and 20,000 vibrations per second. (hertz).

■ **source**

a laser diode or LED used to inject an optical signal into fiber.

■ **source follower**

FET amplifier in which signal is applied between gate and drain with output taken between source and drain. Also called "common drain."

■ **source impedance**

the combination of resistance and reactance that a source presents to the input terminals of a measuring instrument.

■ **source level**

sound pressure (acoustic power) in dB referenced to 1.0 microPascal measured at 1 meter (one foot in air) from the sound source.

■ **source terminal**

along with the gate and drain, one of the three terminals of an FET.

■ **source-measure unit (smu)**

an electronic instruments that sources and measures DC voltage and current. Generally, SMUs have two modes of operation source voltage and measure current, or source current and measure voltage. Programmable features can include constant or sweep source, measurement delays, and compliance levels. Also known as a source/monitor unit or a stimulus/measurement unit.

■ **south pole**

pole of a magnet into which magnetic lines of force are assumed to enter.

■ **space charge**

a cloud of electrons, which gather around a tube electrode (such as the cathode).

s-parameter cable correction

a method of correcting for errors in high frequency capacitance measurements due to cables connecting the device under test (DUT) to the instrument. In this method, the cable impedance is measured by a network analyser and transmitted to the instrument by GPIB. The instrument uses the impedance parameters to correct for cable errors.

■ **spark**

Momentary discharge of electrical energy due to ionisation of air or other dielectric material separating two charges.

■ **SPDT**

single pole double throw.

■ **speaker**

also called "loudspeaker." Transducer that converts electrical energy into mechanical energy at audio frequencies.

■ **spectral map**

a three-dimensional plot of the vibration amplitude spectrum versus another variable, usually time or rpm.

■ **spectral purity**

a description of the distortion components of an oscillator's output signal at a specified amplitude and load. Includes total harmonic distortion, harmonics, and spurious components.

■ spectral response
the response of a device or material to monochromatic light as a function of wavelength.

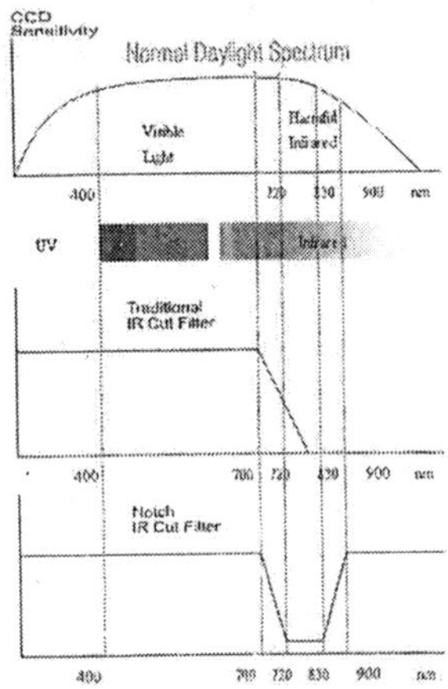

■ spectrum
arrangement or display of light or other forms of electromagnetic radiation separated according to wavelength, energy or some other property.

■ spectrum analyser
instrument used to display the frequency domain of a waveform plotting amplitude against frequency an instrument which displays thefrequency spectrum of an input signal.

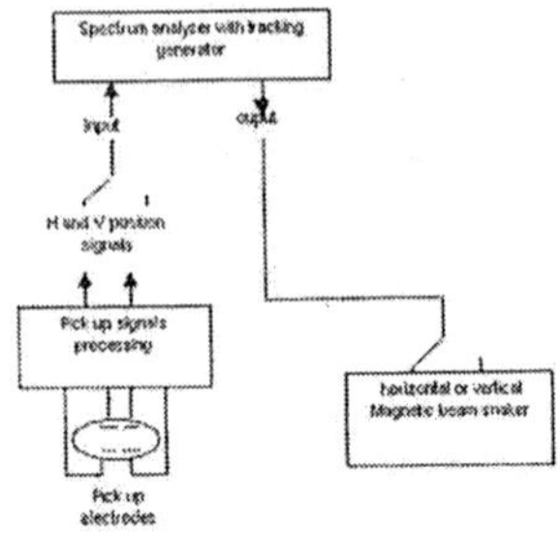

■ speed-up capacitor
capacitor added to the base circuit of a BJT switching circuit to improve the switching time of the device.

■ splice (fusion or mechanical)
a device that provides for a connection between two fibers, typically intended to be permanent.

■ splitting ratio
the distribution of power among the output fibers of a coupler.

■ spreading resistance profiler (srp)
a measurement system that measured carrier concentration by beveling the wafer (grinding it at an angle), then measuring resistance with two probes.

■ SPST
abbreviation for "single pole single throw."

■ spurious components
undesired signals in the output of

a signal generator that have a defined amplitude and frequency. They are not harmonically related to the fundamental frequency. Specified as decibels below the carrier frequency.

spurious-free dynamic range

the ratio of the RMS value of the sine wave f(IN) (input sine wave for an ADC, reconstructed output sine wave for a DAC) to the RMS value of the peak spur observed in the frequency domain. It is typically expressed in decibels. SFDR is important in certain communication applications that require maximizing the dynamic range of the converter. *See also* root mean square (RMS) *and* dynamic range.

square wave

a periodic wave that alternately assumes one of two fixed values (high and low) with the transition time between these two levels being negligible.

SRAM

static Random Access Memory; memory circuits which permit the storage and retrieval of data from any desired memory cell. The data remain stored as long as power is applied.

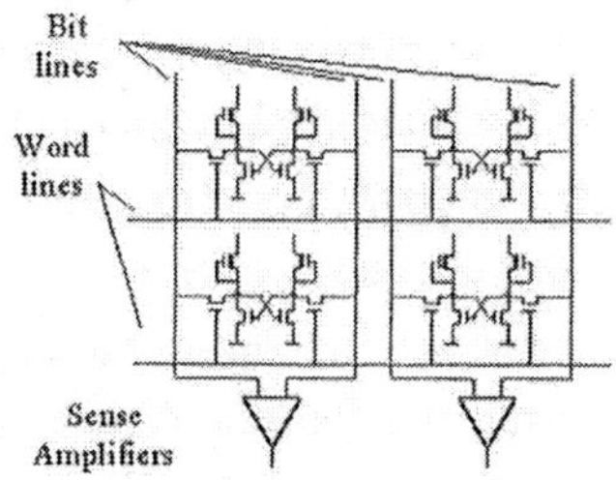

Static Random Access Memory

stability

temperature, electrical, dimensional, and power stability are important attributes in a working laser. The Everlase™ series of industrial lasers achieve such stability through circulating coolant, cast iron end pieces, permanent alignment, electrical control, and kinematic mounting. Result long-term variation of only +/- 2% of rated power.

stage

one of a series of circuits within a single device; e.g., first stage of amplification.

staircase waveform

a waveform in which the voltage is incremented in uniform steps from the start voltage to the stop voltage.

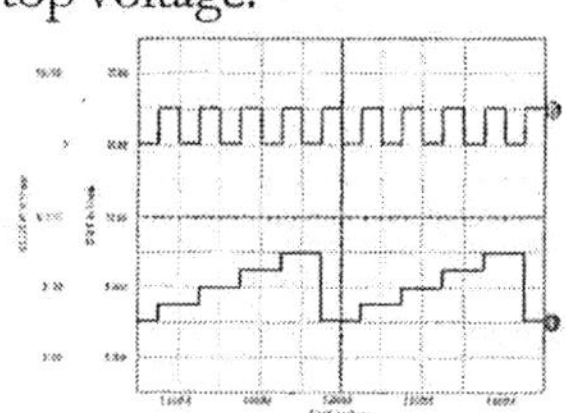

■ standard cell
a pre-designed integrated circuit module that implements a specific function. The data required to use the cell as part of a more complex circuit are contained in a computer file that can be accessed by other CAD tools. Standard cells are designed to be interconnected by simple abutment with other cells from the same cell library.

■ standoff potential
the breakdown voltage across relay contacts.

■ static
crackling noise heard on AM radio receivers. Caused by electric storms or electric devices.

■ static bias current
a current used to put an ASIC in a known operating condition.

■ static electricity
stationary electric charges.

■ static reverse current
reverse current through a zener diode when the reverse voltage across the diode is less than the zener voltage rating of the device.

■ static scanning
scanning method which generates periodic signals during movement; the
signal periods and fractions thereof correspond to a definite linear or
angular displacement.

■ stator
stationary part of some rotary device such as a variable capacitor.

■ steady state modal distribution
equilibrium modal distribution (EMD) in multimode fiber, achieved some distance from the source, where the relative power in the modes becomes stable with increasing distance.

■ step index fiber
a multimode fiber where the core is all the same index of refraction.

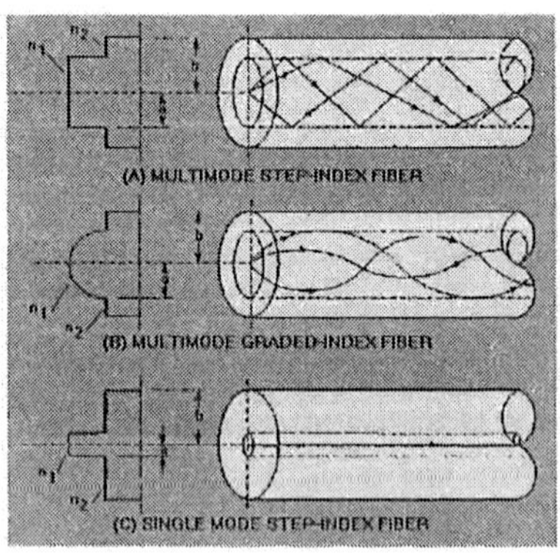

Different Modes of Step Index Fiber

■ step-down transformer

transformer in which the output AC voltage is less than the input AC voltage.

■ step-up transformer

transformer in which the output AC voltage is greater than the input AC voltage.

■ stereo sound

system in which reproduced sound is delivered through two or more channels to give a sense of direction to the source.

■ stereophonic

sound reproduction utilizing separate amplification channels feeding respective loudspeaker systems to provide separation of output signals for various instrumental or vocal groupings.

■ stiffness-the spring

like quality of mechanical and hydraulic elements to elasticity deform under load.

■ stimulated emission

when an atom, ion, or molecule capable of lasing is excited to a higher energy level by an electric charge or other means, it will spontaneously emit a photon as it decays to the normal ground state. If that photonpasses near another atom of the same frequency, which is also at some metastable energy level, the second atom will be stimulated to emit a photon. Both photons will be of the same wavelength, phase, and spatial coherence. Light amplified in this manner is intense, coherent (collimated or parallel), and monochromatic. In short, laser light.

Stimulated Emission in a Mirrored Laser Cavity

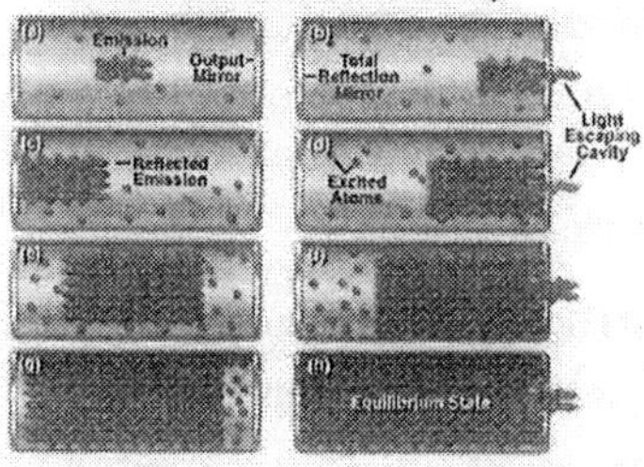

■ stop band

range of frequencies outside the pass band of a tuned amplifier.

■ storage time

in a BJT switching circuit, it is the time required for collector current to drop from 100% to 90% of its maximum value.

■ strain

the physical deformation, deflection, or change in length resulting from stress (force per unit area).

■ strain gauge

a device with electrical resistance that is function of the applied measurand.

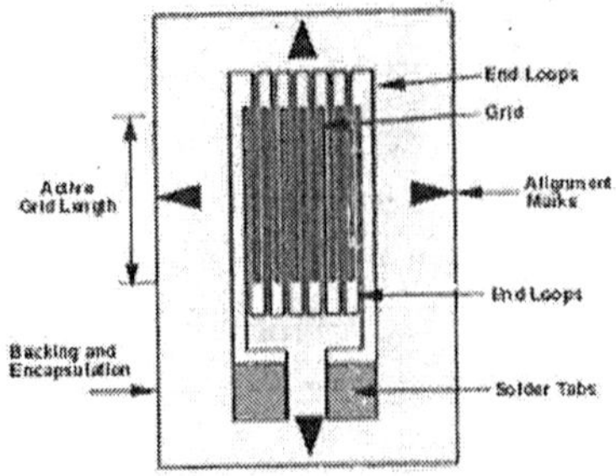

Strain Gauge Internal Structure

■ stranded conductor
conductor composed of a group of strands of wire twisted together.

■ stray capacitance
Undesirable capacitance that exists between two conductors such as two leads or one lead and a metal chassis.

■ stress gradient
stress gradients are produced between layers due to different thermal coefficients of expansion between the layers.

■ stretchout
an effect caused by interface traps, which make the C-V curve take more voltage for the same curve shape. Before quasistatic C-V and simultaneous C-V were predominant, this phenomenon was used to measure Dit using the "Terman" method.

■ subassembly
components contained in a unit for convenience in assembling or servicing equipment.

■ subatomic
Particles such as electrons, protons and neutrons that are smaller than atoms.

■ sub-bottom profiling
determining the sedimentary structure of the ocean floor by utilizing sound waves.

■ subroutine
a set of software instructions executed by a single line of code.

■ substrate
the underlying material upon which a device circuit or epitaxial layer is fabricated. This term also applies to the MIS capacitor, which has only two terminals the gate and substrate.

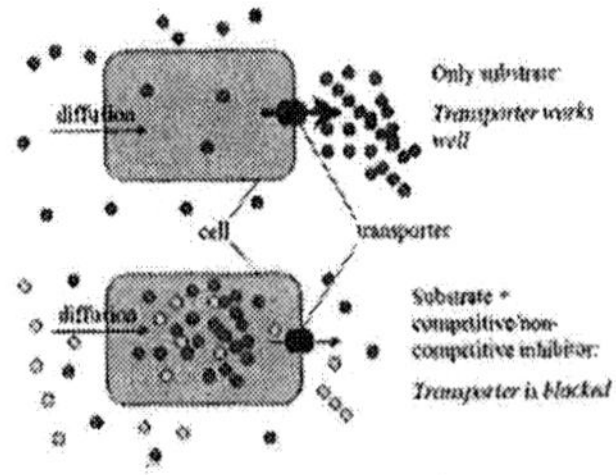

Substrate Prepration

■ sub synchronous
component(s) of a vibration signal whichhas a frequency less than shaft rotative frequency.

■ **sub threshold slopes, linear (parametric testing)**
slope of log (drain current) vs. gate voltage, small drain voltage.

■ **sub threshold slopes, saturation (parametric testing)**
slope of log (drain current) vs. gate voltage, large drain voltage.

■ **subtractor**
a particular type of OPERATIONAL AMPLIFIER where the output voltage is in proportion to the difference between it's inputs. Unlike the COMPARATOR, which can based on exactly the same INTEGRATED CIRCUIT, the Op-Amp has two individual inputs instead of one input and one reference value. This use is sometimes called a DIFFERENTIAL or PROPORTIONAL AMPLIFIER.

■ **summing amplifier**
an op-amp circuit whose output is proportional to the sum of its instantaneous voltages.

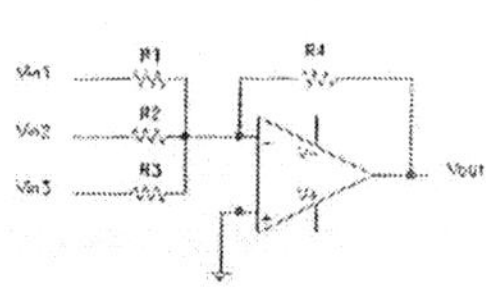

Vout = −R4 • [(Vin1/R1) + (Vin2/R2) + (Vin3/R3)]

■ **super high frequency**
(SHF) Frequency band between 3 GHz and 30 GHz. So desiganted by Federal Communications Comission (FCC).

■ **superconductor**
metal such as lead or niobium that, when cooled to within a few degrees of absolute zero, can conduct current with no resistance.

■ **superheterodyne receiver**
1. The most widely used radio receiver in which the incoming signal is fed into a mixer and mixed with a signal from a local oscillator. The output produces an intermediate frequency that is amplified through an IF strip, detected by a diode and amplified in an audio amplifier.
2. a communications receiver employing a mixer stage to produce an intermediate frequency which is amplified by fixed-selectivity circuits.
3. Radio receiver that converts all radio frequencies to a fixed intermediate frequency to maximise gain and bandwidth before demodulation.

■ **superposition theorem**
theorem designed to simplify networks containing two or more

sources. It states that in a network containing more than one source, the current at any one point is equal to the algebraic sum of the currents produced by each source acting separately.

■ super-regenerative receiver

a receiver employing a detection principle based on injection of a supersonic signal for increased sensitivity.

■ supply voltage

Voltage provided by a power source.

■ suppressor grid

a grid (usually between anode and screen grid) placed at ground potential for the purpose of eliminating secondary emission by redeflecting electrons to the anode.

■ surface emitter led

a LED that emits light perpendicular to the semiconductor chip. Most LEDs used in data communications are surface emitters.

■ surface generation velocity

a parameter derived from C-t using Zerbst analysis. This parameter is related to generation of electron-hole pairs at the semiconductor surface, whereas generation lifetime pertains to generation in the bulk semiconductor material.

■ surface leakage current

diode reverse current that passes along the surface of the semiconductor materials.

■ surface leakages. (parametric testing)

a CMOS or bipolar device parameter or a yield parameter. Leakage per unit surface area.

■ surface potential

for our purposes, this term is equivalent to "band bending."

■ surface-barrier diode

(Schottky diode) high speed diode that has very little junction capacitance. Also known as a "hot-carrier diode."

■ surge current

high charging current that flows into a power supply filter capacitor as the power is first turned on.

■ swamping resistor

a resistor used to increase or "broaden" the bandwidth of a circuit.

■ sweep generator

1. test instrument designed to produce a voltage that continously

varies in frequency over a band of frequencies. Used as a souce to display frequency response of a circuit on an oscilloscope.

2. a signal generator capable of providing frequency variations (sweep) through a predetermined range.

■ swinging choke

a filter inductor employed for voltage-regulation purposes in power supplies. It has saturable-reactor characteristics.

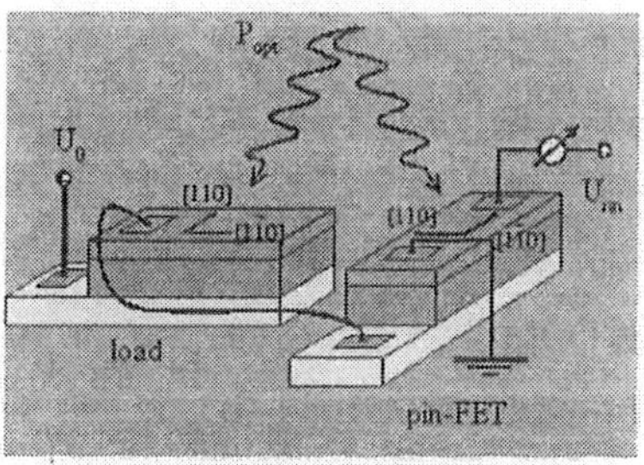

Semiconductor Switch

■ switch card

a type of card with independent and isolated relays for switching inputs and outputs on each channel.

■ switch

Connecting one input to one output. Also known as actuate.

■ switched current

the maximum current level that can be reliably handled while opening and closing contacts.

■ switching mainframe

a switching instrument that operates according to user commands to connect signals among sourcing and measuring instruments and devices under test. a mainframe is also referred to as a scanner, multiplexer, matrix, or programmable switch.

■ switching transistor

transistor designed to change rapidly between saturation and cutoff.

■ SWR

Standing Wave Ratio. Variously VSWR or ISWR for voltage or current SWR. It is the ratio of the voltage (or current) maximum at any point on the line to the voltage (or current) minimum at that point. a perfectly matched system has an SWR of 1. 0:1, meaning that the measurements are the same at any point. Standing waves are generally the result of a mismatch between the antenna and transmission line and form when power is not absorbed by the antenna and is reflected back toward the transmitter. High SWR's can be harmful to the antenna, transmission line and radio equipment.

■ **sync pulse**

pulse used as a reference for synchronisation.

■ **synchronisation**

also called sync. Precise matching of two waves or functions.

■ **synchronous mode**

a standard for data transmission. Data is transferred without start and stop bits but with a clock signal to synchronise the receiver. This gives higher data throughput than asynchronous mode, but can be less secure.

■ **synchronous sampling**

in a DSA, it refers to the control ofthe effective sampling rate of data; which includes the processesof external sampling and computed resampling used in ordertracking.

■ **synchronous**

in hardware, it is an event that occurs in a fixed time relationship to another event. In software, it refers to a function that begins an operation and returns to the calling program only when the operation is complete.

■ **synthesis**

the CAD operation of automatically generating an integrated circuit layout from a high-level description (e. g. using a hardware description language) of the desired circuit function and behavior.

■ **system**

Combination of several pieces of equipment to perform in a particular manner.

■ **systematic error**

reproducible measuring deviation, which can be compensated for by
e.g. computation.

■ **T flip flop**

toggle flip flop.

■ **T pad**

an attenuator whose components are wired to appear as the letter T.

■ **talkset, fiber optic**

a communication device that allows conversation over unused fibers.

■ **tank circuit**

1. parallel resonant circuit containing only a coil and a capacitor. Both the coil and capacitor store electrical energy for part of each cycle.

2. usually refers to the anode resonant circuit (L and C in parallel); it also applies to any parallel resonant circuit storing r-f energy by virtue of its fly-wheel effect.

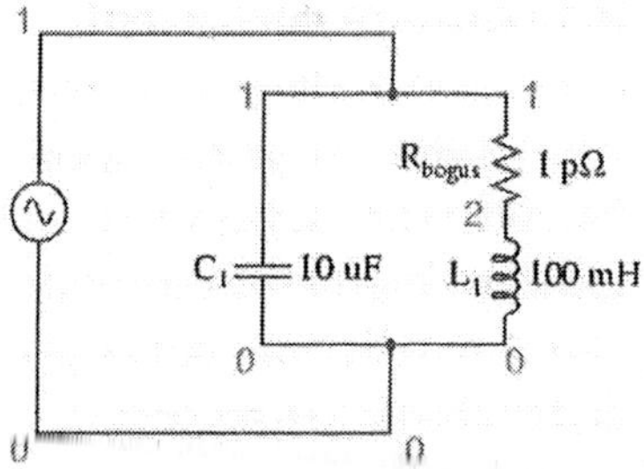

Simple Parallel Resonance

■ **tantalum capacitor**
electrolytic capacitor having a tantalum foil anode. Able to have a large capacity in a small package.

■ **tap**
electrical connection to some point other than at the ends of a resistor or inductor.

■ **tapered**
nonunifrom distribution of resistance per unit length throughout the element of a potentiometer.

■ **target strength**
a measure of the percentage of the acoustic energy hitting the target that is reflected back to the transducer.

■ **technician**
expert in troubleshooting circuit and system malfunctions. Along with a thorough knowledge of test equipment and how to use it to diagnose problems, the technician is also familiar with how to repair or replace faulty components. Technicians basically translate theory into action.

■ **telegraphy**
communication between two points by sending and receiving a series of current pulses either through wire or by radio.

■ **telemetry**
transmission of instrument readings to a remote location either by wire or by radio.

■ **telephone**
apparatus designed to convert sound waves into electrical waves which are sent to and reproduced ata distant point.

■ **telephone line**
wires existing between subscribers and central stations in a telephone system.

■ **telephony**
telecommunications system involving the transmission of speech information, allowing two or more persons to communicate verbally.

■ **teletypewriter**
electric typewriter that like a teleprinter can produce coded signals corresponding to the keys

pressed or print characters corresponding to the coded signals received.

television
system that converts both audio and visual information into corresponding electrical signals which are then transmitted through wires or by radio waves to a receiver which reproduces the original information.

telex
teletypewriter exchange service.

TEM
abbreviation for Transverse Electromagnetic Mode, the cross-sectional shape of the working laser beam. an infinite number of shapes can be produced, but only a relatively small number are needed for industrial applications. In general, "the higher the TEM, the coarser the focusing."

temperature coefficient of frequency
rate at which frequency changes with temperature.

temperature coefficient of resistance
the amount of resistance change of a material per degree of temperature change.

temperature coefficient
a change in reading (or sourced value) with a change in temperature. it is expressed as a percentage of reading (or sourced value), plus a number of counts per degree change in temperature.

temperature effects
(On rated output) The change ofrated output for a specified temerature change at steady statetemerature conditions. Expressed as a percentage of theoutput per °C. (On zero output) The change of zero outputfor a specified temerature change at steady state temperatureconditions. Expressed as a percentage of the rated output per °C.

temperature range-compensated
the temperaturerange over which the sensor will meet its stated specification.

tera
(T) metric prefix that represents 1012.

terminal
point at which electrical connections are made.

termination
preparation of the end of a fi-

ber to allow connection to another fiber or an active device, sometimes also called "connectorisation".

tesla

(T) unit of magnetic flux density. (1 tesla = 1 Wb/m^2).

test

sequence of operations intended to verify the correct operation or malfunctioning of a piece of equipment or system.

test cable

a short single fiber jumper cable with connectors on both ends used for testing. This cable must be made of fiber and connectors of a matching type to the cables to be tested.

test kit

a kit of fiber optic instruments, typically including a power meter, source and test accessories used for measuring loss and power.

test source

a laser diode or LED used to inject an optical signal into fiber for testing loss of the fiber or other components.

test structure

a device manufactured on a wafer expressly for the purpose of testing.

tetrode

a four-electrode tube or transistor.

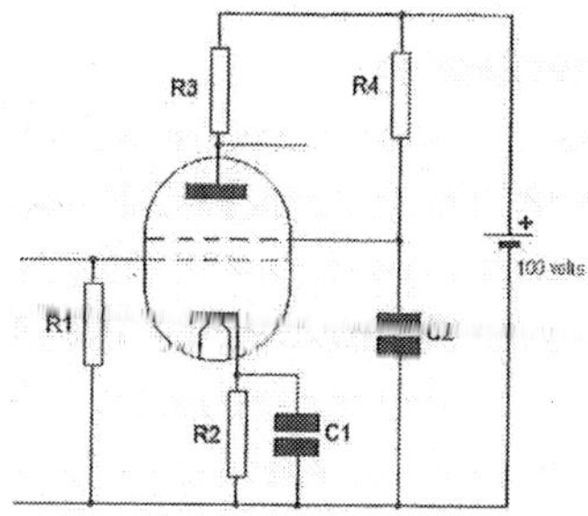

A Tetrode Vacuum Tube

thermal relay

relay activated by a heating element.

thermal runaway

problem that can develop in an amplifier when an increase in temperature causes an increase in collector current. The increase in collector current causes a further increase in temperature and so on. Unless the circuit is designed to prevent this condition, the device can be driven into saturation.

thermal shutdowns

a device used to shut down circuits before they reach a self-destructive temperature.

thermal stability

the ability of a circuit to maintain stable characteristics in spite of increased temperature.

■ thermionic emission

the production of electrion emission by thermal means.

■ thermistor

this is a special type of RESISTOR. The resistance of all resistors changes slightly as they get hotter. Thermistors are usually made so that their resistance falls noticeably as the temperature around them rises. They can be used to detect temperatures very accurately when used with a simple COMPARATOR circuit. The resistance of a thermistor is specified at a certain temperature, for instance 47K at 25C. At low temperature they have a high resistance. High temperature they have a low resistance. They are input devices.

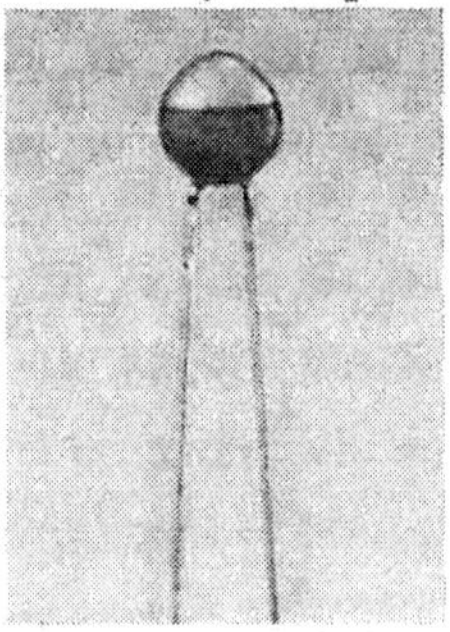

■ thermocouple

a temperature sensor created by joining two dissimilar metals. This junction creates a small voltage as a function of the temperature.

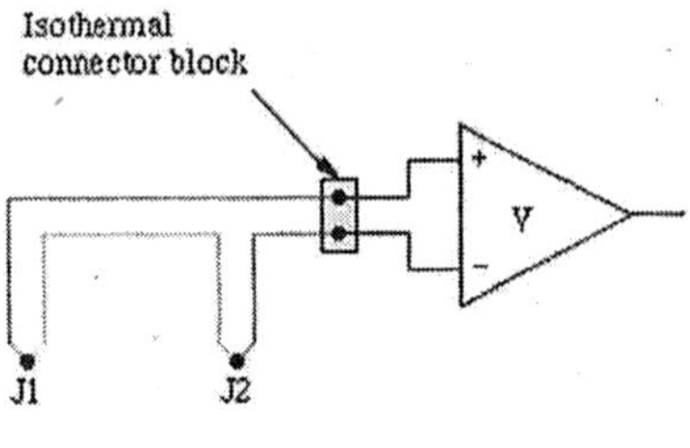

An Electronic Thermocouple

■ thermoelectric voltage

Voltages resulting from temperature differences within a measuring circuit or when conductors of dissimilar materials are joined together. Also known as thermal EMF or thermal offset.

■ thermometry

relating to the measuring of temperature.

■ thermostat

device that opens or closes a circuit in response to changes in temperature.

■ Thevenin's theorem

theorem that replaces any complex network with a single voltage source in series with a single resistance.

■ thick film resistor

fixed value resistor consisting of thick-film resistive element made from metal particles and glass powder.

■ **thin film capacitor**

capacitor in which both the electrodes and the dialectric are deposited in layers on a substrate.

■ **thin film detector**

(TFD) a temperature detector containing a thin layer of platinum and used for precise temperature readings.

■ **thin film**

a thin layer of material grown or deposited on a substrate. Typically, layer thickness is measured in tens to thousands of angstroms for insulators and microns for metals.

■ **three phase supply**

aC supply that consists of three AC voltages 120° out of phase with each other.

■ **threshold voltage**

the gate voltage needed to turn on a MOS enhancement-mode device. Minimum point at which an effect is produced or detected.

■ **threshold voltages. (parametric testing)**

a CMOS device parameter. Voltage at 40 nAW/L, or extrapolation from maximum linear slope.

■ **throughput**

the maximum rate at which a data conversion system can perform repetitive conversions within a specified accuracy. It is determined by summing the various times required for each part of the conversion system and then taking the inverse of this time. The throughput rate takes into account the total time required to process a signal and store the value in either onboard or system memory.

■ **thyratron**

a gas-tube rectifier with a grid electrode that has the ability to initiate current flow when the anode is positive. Once conduction occurs, the grid loses control until the plate current is interrupted.

■ **thyristor**

1. a term used to classify all four layer semiconductot devices. SCRs and triacs are examples of thyristors.

2. a component rather like a diode but will not conduct until a voltage is applied to its third terminal known as the gate.

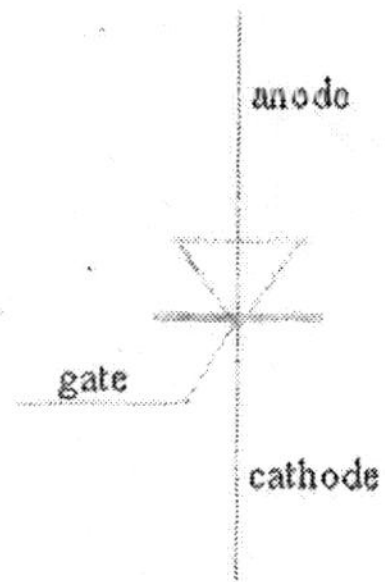

■ time averaging
in a DSA, averaging of time records that results in reduction of asynchronous components with reference to the trigger.

■ time constant
(.) Time required for a capacitor in an RC circuit to charge to 63% of the remaining potential across the circuit. Also time required for current to reach 63% of maximum value in an RL circuit. Time constant of an RC circuit is the product of R and C. Time constant of an RL circuit is equal to inductance divided by resistance.

■ time division multiplexing
(TDM) transmission of two or more signals on the same path, but at different times.

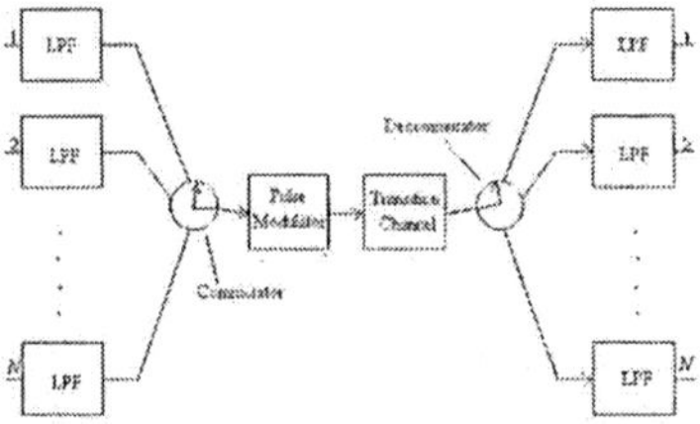

Schematic of the Division Multiplexing

■ time record
in a DSA, the sampled time data converted tothe frequency domain by the FFT. Most DSAs use a time recordof 1024 samples.

■ time base accuracy
a measure of how closely the internal timebase of an instrument tracks a known time standard.

■ time-domain analysis
a method of representing a waveform by plotting amplitude over time.

■ timing jitter
the short-term variation of the time period between sample points.

■ tLV-STEL
threshold Limit Value-Short-Term Exposure Limit is a 15 minute time weighted average exposure that should not be exceeded at any time during a work day, even if the 8 hour TWa is within the TLV. Exposures at the STEL concentration should not be repeated more than 4 times a day and there should be at least 60 minutes between successive exposures at the STEL.

■ TLV-TWA
threshold Limit Value-Time Weighted Average is the time weighted average concentration for a normal 8 hour work day and a 40 hour work week, to which nearly all workers may be exposed without adverse effect.

■ **TO package**

cylindrical, metal can type of package of some semiconductor components.

■ **toggle switch**

spring loaded switch that is put in one of two positions either on or off.

■ **tolerance**

the allowable percentage variation of any component from that stated on its body. For instance red=2%, gold=5% and silver=10%.

■ **toroidal coil**

coil wound on a doughnut shaped core.

■ **torsional stiffness**

rotational rigidity of a precision coupling governing the reversal error of a rotary encoder.

■ **torsional vibration**

amplitude modulation of torque measuredin degrees peak-to-peak referenced to the axis of shaft rotation.

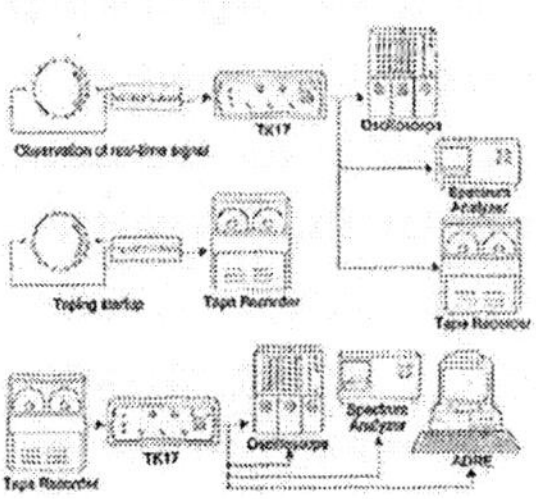

Set up of Torsional Vibration

■ **total harmonic distortion**

1. the rms value of the distortion appearing at multiples (harmonics) of the input (or output for a DAC) frequency to the RMS value of the input (or output) sine wave. Only harmonics within the Nyquist limit are included in the measurement. It is typically expressed in decibels. *See also* root mean square (RMS). The total amount of distortion over a specified frequency range an analog circuit introduces after processing a signal.

2. the percentage of harmonic distortion present in an output signal over a specified frequency range

■ **total internal reflection**

confinement of light into the core of a fiber by the reflection off the core-cladding boundary.

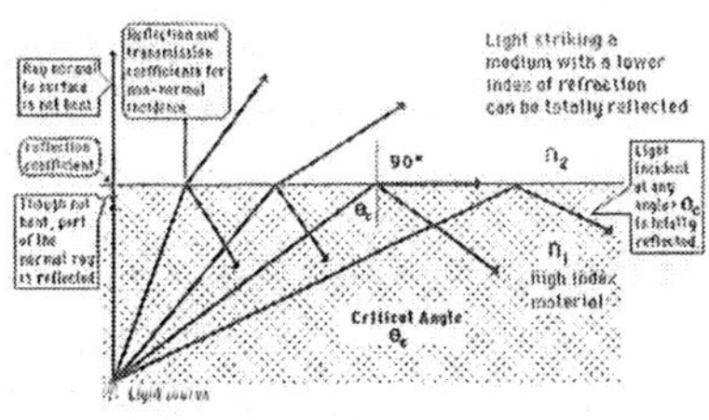

■ **tracking**

the condition which prevails when simultaneously tuned circuits maintain close resonance for the tuned signal frequency.

tracking filter

a low-pass or band-pass filter which automatically tracks the input signal versus the rpm. a tracking filter is usually required for aliasing protection when data sampling is controlled externally.

transducer

1. a device than converts some physical condition such as temperature, pressure, humidity, etc. into an electrical signal (typically a voltage or current). The output generally needs signal conditioning because it contains errors (offset, non-linearity, temperature dependence), has too small an amplitude, or is too weak to drive a load.

2..in acoustics this term is used to describe an antenna which converts electrical energy into sound wave and vice versa.

3.a device for translating the magnitude of one quantity into another quantity.

transfer accuracy

a comparison of two nearly equal measurements over a limited temperature and time. It is expressed in ppm.

transfer rate

the rate at which data is transferred to or from memory.

transformation hardening

a process well-suited to lasers which involves hardening metal by heating it to the critical temperature for transformation, quenching it, and solidifying it with uniform distribution of its carbon content.

transformer

inductor with two or more windings. Through mutual inductance, current in one winding called a primary will induce current into the other windings called secondary.

transformer coupling

also called inductive coupling. Coupling of two circuits by means of mutual inductance provided by a transformer.

transient

a signal component of fractional duration as compared to the primary signal.

transient response

the time-domain response observed when a signal is applied to a circuit. The response is often measured by the amount of time it takes for an analog signal to reach and stay within a specified error band from an instantaneous change in input conditions.

transient vibration

temporarily sustained vibration of a mechanical system. It may consist of forced or free vibration or both. Typically this is associated with changes in machine operating condition such as speed, load, etc.

transistor

1. a three leaded device (Collector, Base, Emitter) used for amplifying or switching. Also called a bi-polar transistor to distinguish it from Field Effect Transistor etc.

2. a semiconductor device in which a small control signal is used to control a larger current flow.

3.a transistor is a SOLID STATE switch and is cheaper than a RELAY. The current being controlled enters at the COLLECTOR, but will not go any further until the GATE opens, after which it leaves at the EMITTER. Many different types of transistor exist, including special types like PHOTOTRANSISTORS, but the most common are 'NPN bi-polar' types. a transistor can amplify an output signal, often by several hundred times. This is called GAIN .Transistors can act as a switch or an amplifier. They are passive components .

4. a solid-state device having amplifying, switching, and signal-generating characteristics.

transistor array

multiple discrete (usually unconnected) transistors on a single silicon chip.

transistor breakdown (parametric testing)

a bipolar device parameter. primarily collector to emitter breakdown, base open.

transition noise (referring to the ADC)

the range of input voltages that cause the analog-to-digital converter (ADC) result to toggle between codes. As the input voltage is increased, the voltage that defines where a code transition occurs (code edge) has an associated amount of noise due to this transition. This specification is often given as an RMS value rather than peak-to-peak.

■ **transmission**

1. sending of information.

2. in optics, the passage of radiant energy (light) through a medium.

■ **transmission line**

conducting line used to transmit signal energy between two points.

■ **transmit current response**

the level of the acoustic output referenced to one meter (one foot in air) per one amp input.

■ **transmit voltage response**

the level of the acoustic output referenced to one meter (one foot in air) per one volt input.

■ **transmittance**

the ratio of transmitted radiant energy to incident radiant energy, or the fraction of light that passes through a medium.

■ **transmitter (Tx)**

1. a device that converts audio, video or coded signals into modulated radio frequency signals which can be propagated by electromagnetic waves (radio waves).

2. a device which includes a LED or laser source and signal conditioning electronics that is used to inject a signal into fiber.

■ **transponder (acoustic)**

a devise that automatically transmits sonar signals when actuated by a specific sonar signal from an interrogator. Transponders are used to mark or track objects or sites underwater. They are programmed to be in a continuous passive (listening) mode until they receive a valid signal from a transponder interrogator.

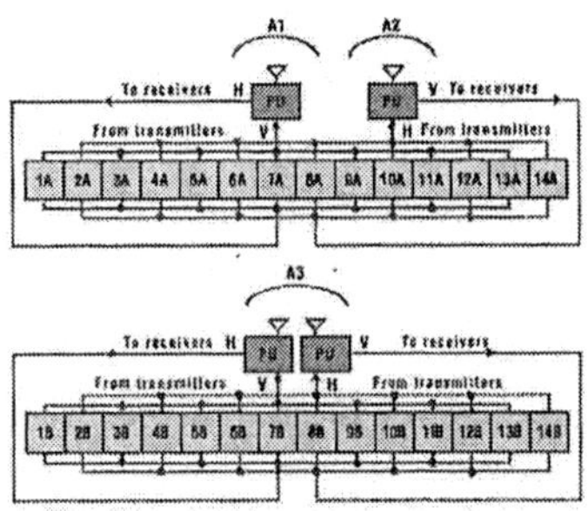

■ **trepanning the beam**

relative motion of the beam with respect to the part, usually in a circular fashion.

■ **triac**

bidirectional gate controlled thyristor similar to an SCR, but capable of conducting in both directions. Provides full wave control of AC power. Similar to a thyristor but allows current to flow in both directions as in diagram below.

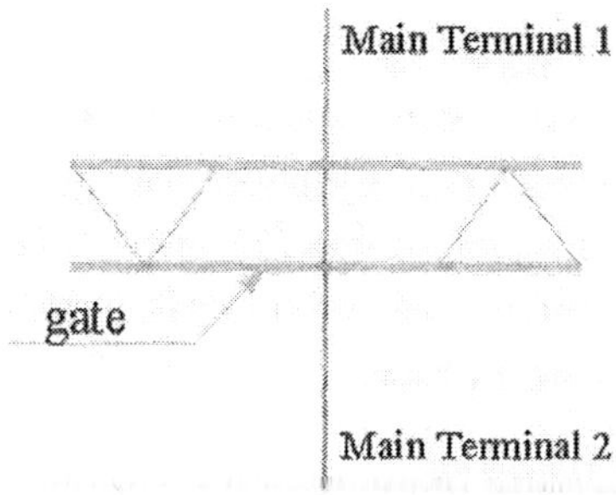

■ triangular wave

a repeating wave that has equal positive going and negative going ramps. The ramps have linear rates of change with time.

■ trigger

1. any event which can be used as a timing reference. In a DSA, a trigger can be used to initiate a measurement.

2. pulse used to initiate a circuit action.

■ trigger (data acquisition)

an event that starts or stops an operation. a trigger can be a specific analog, digital, or software condition.

■ trigger (instrumentation)

an external stimulus that initiates one or more instrument functions. Trigger stimuli include an input signal, the front panel, an external trigger pulse, and IEEE-488 bus X, Talk, and GET triggers.

■ trigger conditions

refers to trigger sensitivity, polarity, etc.

■ trigger hysteresis

applies only to analog triggers. It helps prevent noise from triggering an operation or event. For a positive-edge trigger, the analog signal must be above the specified voltage level by at least the amount of the hysteresis value before the trigger occurs. For a negative-edge trigger, the analog signal must be below the specified voltage level by at least the amount of the hysteresis value before the trigger can occur.

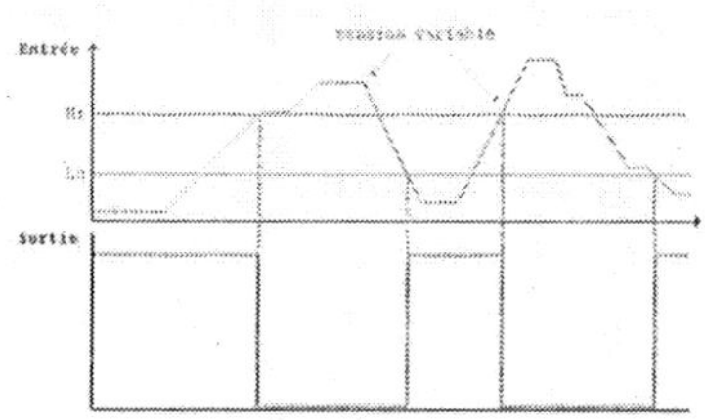

Trigger Hysteresis Curve

■ trigger jitter

the short-term variation of the time period between a trigger event and the first sample point.

■ trigger latency

the fixed time offset between the trigger event and the first sample point.

■ **trigger mode**
refers to when data acquisition begins and ends in relationship to the trigger. Trigger modes include normal-trigger, pre-trigger, about-trigger, post-trigger, trigger-to-trigger, and trigger-to-about-trigger.

■ **trigger polarity**
for edge-sensitive triggers trigger polarity defines whether the trigger occurs when the signal is rising (positive direction) or falling (negative direction). For level-sensitive triggers trigger polarity defines whether the trigger occurs when the signal is above a level (positive) or below a level (negative).

■ **trigger sensitivity**
refers to edge and/or level of a trigger. For analog triggers, trigger sensitivity defines whether the trigger occurs on a transition across a specified value (edge) or whether the trigger occurs when it is above or below a specified value (level). For digital triggers, trigger sensitivity defines whether the trigger occurs on a transition from one state to another state (edge) or whether the trigger occurs when it is at a specified value (level).

■ **triggering**
initiation of an action in a circuit which then functions for a predetermined time. Example: The duration of one sweep in a cathode ray tube.

■ **trimmer**
small value variable capacitor, resistor or inductor used to fine tune a larger value.

■ **trimmer**
trimmer Capacitor. a small variable capacitor used in paralled across a larger capacitor to adjust the total capacitance a small amount.

■ **trivalent element**
one having three valence electrons. Used as an impurity in semiconductor material to produce p-type material. Most commonly used trivalent elements are: Aluminium, Gallium and Boron.

■ **troubleshooting**
systematic approach to locating the cause of a fault in an electronic circuit or system.

■ **true power**
in ac, the product $EI \cos \phi$ (the apparent power multiplied by the power factor).

■ **truth table**
a tabular representation of a logic

gate showing all the possible combSinations.

■ **TTL (or cmos, ecl, etc.) comparator**

a voltage comparator whose high and low output levels match the requirements for TTL (or CMOS, ECL, etc.) logic family input levels.

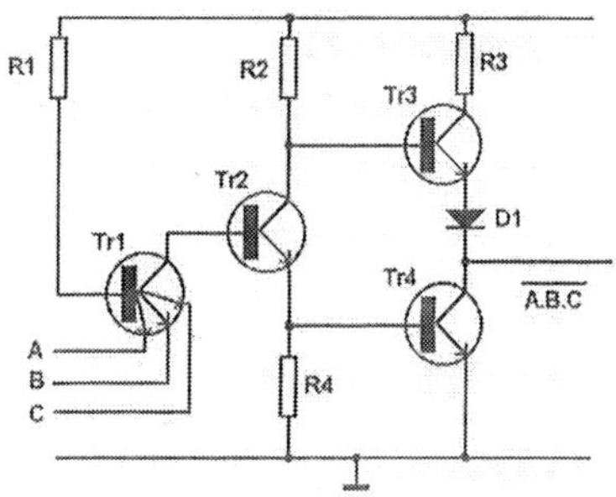

3 INPUT NAND

TTL Circuit Equivalent to 3-Input Nand

■ **TTL**

abbreviation for transistor-transistor-logic. a popular logic circuit family that uses multiple-emitter transistors. a low signal state is defined as a signal 0.8V and below. a high signal state is defined as a signal +2.0V and above.

■ **tuned circuit**

1. a circuit in resonance at a particular frequency.

2. an LC circuit used as frequency-determining network.

3. circuit that can have its component values adjusted so that it responds to one selected frequency and rejects all others.

■ **tuned radio frequency (TRF)**

an amplitude modulated (AM) receiver with one or more stages of radio frequency before the detector.

■ **tuning**

to adjust in a systematic way. Tuning a filter means adjusting components (e.g., R, C, L) to the right values to meet the desired filter response.

■ **tunnel diode**

heavily doped junction diode that has negative resistance in the forward direction of its operating range.

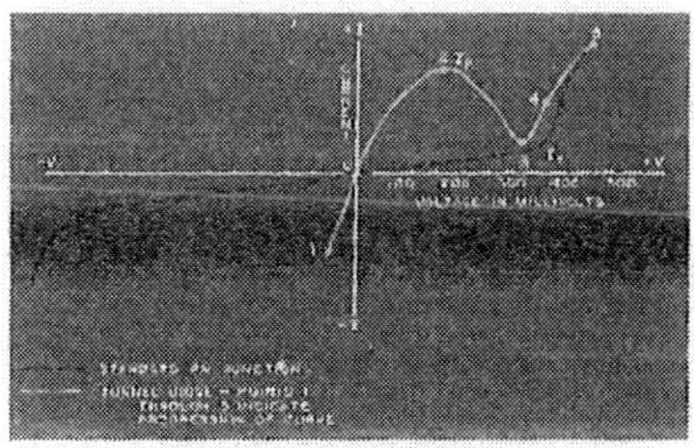

V-I Response a Tunnel Diode

■ **tunneling current**

tunneling current is classically forbidden current across an insulator that is predicted by the quantum mechanics model.

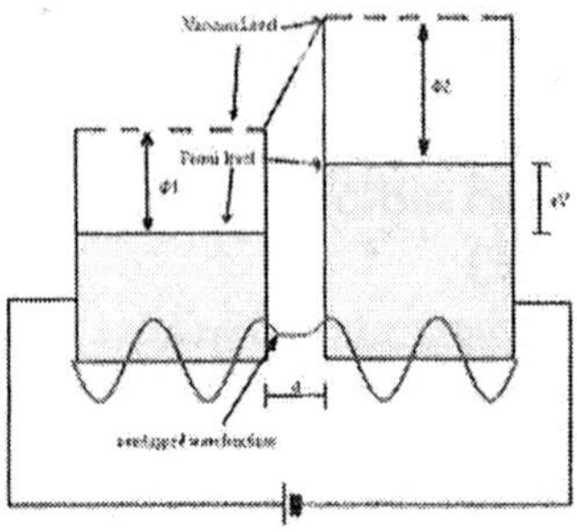

■ **turner test structures**
structures that are specifically designed for high speed process monitoring of product reliability factors. Each structure is tailored to conform to a fab's specific process and design rules.

■ **turn-off time**
sum of storage time and fall time.

■ **turns ratio**
ratio of the number of turns in the secondary winding of a transformer to the number of turns in the primary winding.

■ **tweaking**
to change slightly a signal in a circuit to compensate for some variation due to component or system inaccuracies. This is often done to finalise a system's settings in order to optimise its operation.

■ **two phase**
two repeating waveforms having a phase difference of 90°.

■ **two's complement coding**
a binary digital coding scheme for negative numbers that simplifies addition and subtraction computations. In this scheme, the number "-2" is represented by 11111110 and "2" is represented by 00000010.

■ **two-terminal resistance measurement**
a measurement where the same current flows through the unknown and the test leads.

■ **UHF**
Ultra High Frequency.

■ **ultrasonic**
1. signals that are just above the frequency range of human hearing of approximately 20 kHz.
2. a frequency above the limits of the human ear - above 20kHz.

■ **UMTS**
Universal Mobile Telecommunications System; the name for the "third generation" mobile telephone standard in Europe, standardised by ETSI (European Telecommunications Standardisation Institute).

■ **uncharged**
material having atoms with the same number of electrons in orbit as the number of protons in

the nucleus. Having no electrical charge.

under/over voltage lockouts

a device used to shut down operations before malfunction due to excessive or insufficient voltage.

undersampling

a technique whereby the analog-to-digital converter (ADC) sampling rate is lower than the input frequency (which normally results in a loss of signal information), which causes aliasing. With proper filtering of the input signal and proper frequency selection (including fIN and fSAMPLE), the aliased components that contain the signal information can be shifted from a higher frequency to a lower frequency and converted. This method effectively uses the ADC as a downconverter by shifting higher bandwidth signals into the ADC's desired band of interest. The bandwidth of the ADC's track/hold must be capable of handling the highest frequency signals for this technique to work. *See also* aliasing.

uniform window

in a DSA, a window function with uniformweighting across the time record. This window does not protectagainst leakage, and should be used only with transient signalscontained completely within the time record.

unijunction transistor

1. three terminal device that acts as a diode with its own internal voltage divider biasing circuit.
2. a trigger device that has an emitter lead (e) and two bases B1 and B2.

unipolar

a signal that swings from zero to the positive full-scale range, thus having only a positive polarity.

unity gain frequency

frequency of operation for a device where the gain of the component drops to unity.

universal output transformer

usually refers to an audio output transformer having a number of taps on the secondary winding which provide a variety of impedances for matching the speaker impedance to that of the recommended load resistance.

unmodulated

an r-f carrier signal without modulation.

■ **VA**
abbreviation for "volt ampere"

■ **vacuum tube**
electron tube evacuated to such a degree that its electrical characteristics are essentially unaffected by the presence of residual gas or vapour. Have been essentially replaced by transistors for amplification and rectification. Cathode ray tubes are still used as display devices.

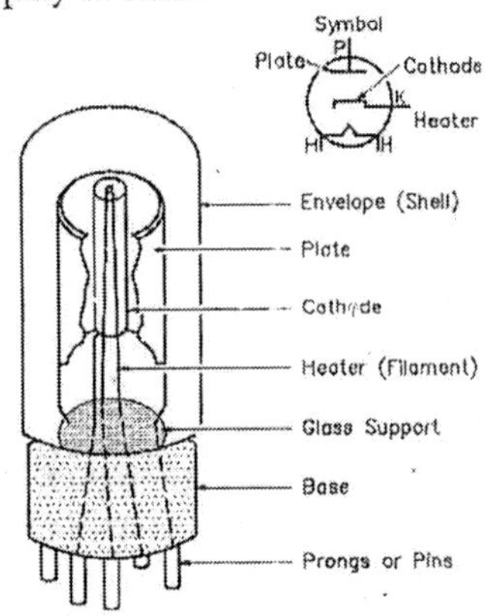

■ **vacuum-tube voltmeter**
a high input-impedance test instrument designed around a tube (or transistor) circuit.

■ **valence band**
the energy levels of the electrons in he outermost shell of the atoms making up a solid material.

■ **valence shell**
the outermost electron shell for a given atom. The number of electrons in this shell determines the conductivity of the atom.

■ **vapour density**
the density of a vapour compared to the density of an equal amount of air.

■ **vapour**
a gaseous form of a material that is normally a solid or liquid at room temperature and pressure.

■ **varactor diode**
PN junction diode with a high junction capacitance when reverse biased. Most often used as a voltage controlled capacitor. The varactor is also called: varicap, tuning diode and epicap.

■ **variable capacitor**
capacitor whose capacitance can be change by varying the effective area of the plates or the distance between the plates.

■ **variable frequency**
frequency settable over a range. This term is applied to a capacitance meter that offers a choice of test signal frequencies from so many that it is essentially continuously adjustable.

■ **variable gain amplifier (VGA)**
an amplifier whose gain can be varied by an external analog or

digital control signal. a VGa can be implemented with an OTa.

■ **variable resistor**

1. resistor whose resistance can be changed by turning a shaft. See also "potentiometer and rheostat."

2. this is an adjustable RESISTOR. They can normally be set to give anything from zero resistance to whatever is the maximum design value. Most have three pins so that they can also be used as a VOLTAGE DIVIDER. By turning the spindle the resistance between the middle pin and the outside pins varies. a variable resistor is an input device.

■ **variable-frequency oscillator**

a signal-generating circuit where component values may be changed for altering the frequency of the output signal.

■ **VCR**

abbreviation for "video cassette recorder."

■ **VCSEL**

vertical cavity surface emitting laser, a type of laser that emits light vertically out of the chip, not out the edge.

■ **vector**

1. quantity having both magnitude and direction. Normally represented by a line. Length of the line indicates magnitude and orientation indicates direction.

2. a quantity which has both magnitude and direction (phase).

■ **vector diagram**

arrangement of vectors showing phase relationships between two or more AC quantities of the same frequency.

■ **vero**

veroboard or strip board or copper strip board is a type of plastic board in ehich there are a series of predrilled holes. The holes are joined together in rows. The componets are pushed through the holes an soldered to the strips. Jumps across the strips can be made with pieces of wire.

■ **vertical MOS**

enhancement type MOSFET designed to handle much greater values of drain current than standard E-MOSFET.

■ **very high frequency**

(VHF) electromagnetic frequency band from 30 MHz to 300 MHz.

■ **very low frequency**

(VLF) frequency band from 3

kHz to 30 kHz.

VHDL

Very high-speed integrated circuit Hardware Description Language; a standardized example of a hardware description language that has the capability of specifying function and behavior of circuits.

via

a plated hole used to provide low inductance connections on an IC. a via may be formed through the dielectric layer on the top of a slice to interconnect two metallisation patterns or may be formed through the back of a slice containing FETs where a low inductance earth is required.

video

relating to any picture or visual information. From the latin word meaning "I see."

video amplifier

1. amplifier having one or mare stages designed to amplify video signals.
2. an amplifier designed to amplify the entire band of frequencies from 10 hertz (10 Hz) to six megahertz (6 MHz). Also called a WIDE-BAND AMPLIFIER.

video crosspoint switch

a crosspoint that has the high level of performance required for video applications.

vignetting

the loss of light through an optical element when the entire bundle does not pass through; an image or picture that shades off gradually into the background.

villari effect

the change in magnetic properties of a ferromagneticmaterial in response to the presence of stress in the ferromagnetic material.

virtual ground

1. point in a circuit that is always at approximately ground potential. Often a ground for voltage, but not for current.
2. a point in a circuit which is at ground potential (0 V) but is not connected to ground.
3. a point in a circuit which is at ground potential (0 V) but is not connected to ground.

visible light transmission/transmittance.

4. the amount of visible light usable to the eye that passes through a filter. As a rule of thumb, as optical density increases, visible light transmission decreases — but not always.

visual fault locator

adevice that couples visible light into the fiber to allow visual tracing and testing of continuity. Some are bright enough to allow finding breaks in fiber through the cable jacket.

VLSI

Very Large Scale Integration; a somewhat generic term applied to most modern integrated circuits which comprise from hundreds of thousands up to millions of individual components. Also applied to systems that employ VLSI circuits.

voice coil

coil attatched to the diaphram of a moving coil loudspeaker. The coil is moved through an air gap between magnetic pole pieces.

voice synthesiser

synthesiser that can simulate speech by stringing together phonemes.

volt

unit of potential difference or electromotive force. One volt is the potential difference needed to produce one ampere of current through a resistance of one ohm. . One thousand volts = 1kV.

voltage

(V) term used to designate electrical pressure or force that causes current to flow.

voltage amplifier

amplifier designed to build up signal voltage. By design amplifiers can have a large voltage gain or a large current gain or a large power gain. Voltage amplifiers are designed to maximise voltage gain often at the expense of current gain or power gain.

voltage burden

the voltage drop across the resistor for the resistance insertion technique of current measurement.

voltage comparator

an amplifier that compares the magnitude of the voltages at its 2 inputs, and outputs a "high" level if +IN > -IN, or a "low" level if +IN < -IN. Basically a 1-bit ADC.

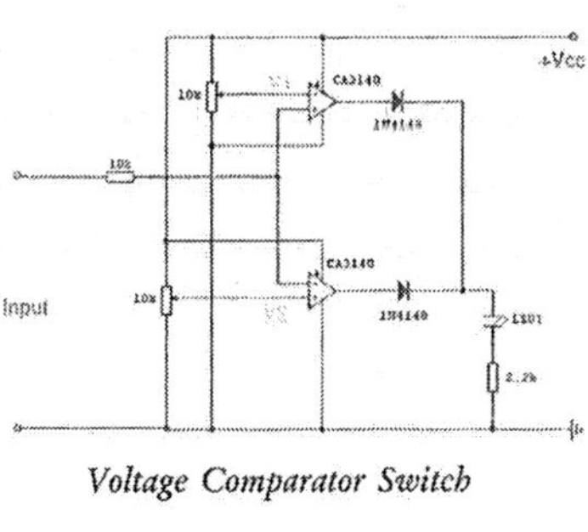

Voltage Comparator Switch

■ voltage controlled oscillator (VCO)

1. a circuit whose output is an AC signal (square wave or sine wave) at a fixed frequency (oscillation) determined by the voltage on a control input. Changing the voltage changes the output frequency. 2. an oscillator whose output frequency varies according to an input signal.

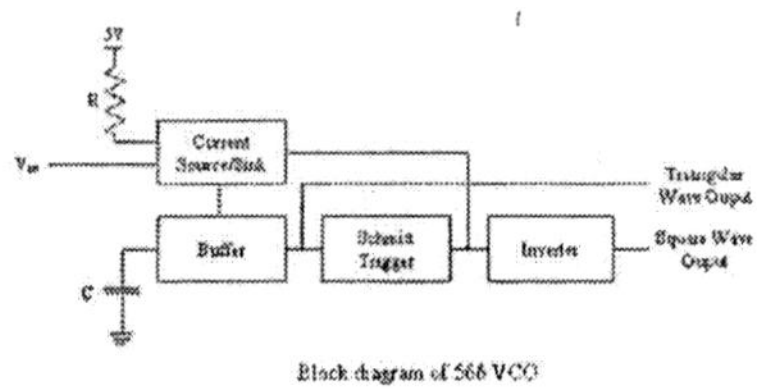

Block Diagram Voltage Controlled Oscillator

■ voltage divider

by connecting two or more resistors in SERIES between the positive and negative supplies, you can split the voltage down into a series of steps. The sise of these steps depends on the resistor values used. For instance, if two resistors, one 220 and another 470, were connected in series in that order between a 9 VOLT supply and GROUND, you would find that the voltage at the point in between the resistors is 6.13V. The first 220 resistor loses 2.87V across it. The calculation is based on the ratio of sises of the resistors used. VARIABLE RESISTORS can be use - in this case, the middle pin gives the divided voltage, normally any value in between the voltage difference across the other two pins.

■ voltage divider biasing

biasing method used with amplifiers in which two series resistors connected across a source. The junction of the two biasing resistors provides correct bias voltage for the amplifier.

■ voltage drop

voltage or difference in potential developed across a component due to current flow.

■ voltage feedback

feedback configuration where a portion of the output voltage is fed back to the input of an amplifier.

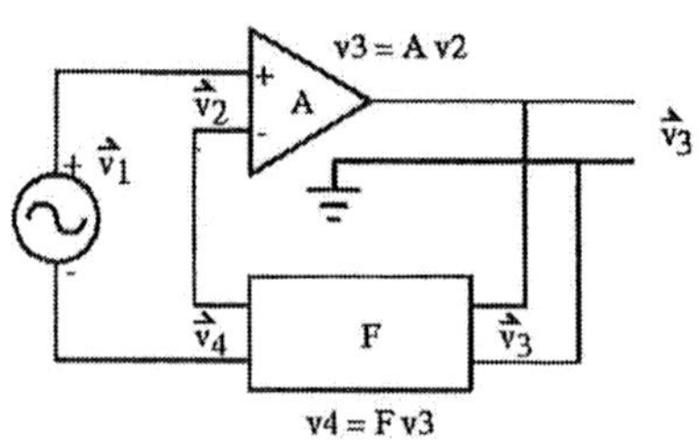

■ voltage feedback amplifier (VFA)

the original op amp architecture, especially useful for low noise and precision applications. Nearly all of the Ca and Ha type op amps are VFAs.

■ **voltage follower**

1. operational amplifier circuit characterized by a high input impedance, low output impedance and unity voltage gain. Used as a buffer between a source and a low impedance load.
2. an op amp connected in a gain of 1 (unity gain).

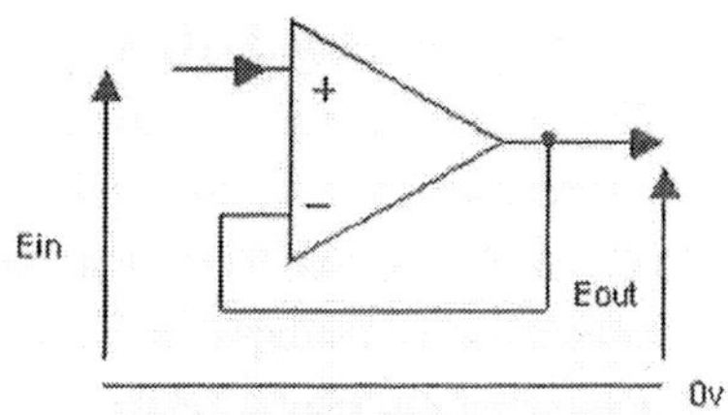

■ **voltage gain**

also called voltage amplification. Ratio of amplifier output voltage to input voltage usually expressed in decibels.

■ **voltage multiplier**

1. rectifier circuit using diodes and capacitors to produce a DC output voltage that is some multiple of the peak value of AC input voltage. Cost effective way of producing higher DC voltages. Voltage doublers and voltage triplers are examples.
2. a set of components designed to increase the voltage for a particular application.

■ **voltage rating**

maximum voltage a component can withstand without breaking down.

■ **voltage reference**

a circuit that produces a very precise and stable voltage over a range of temperature and power supply conditions.

■ **voltage regulator**

device or circuit that maintains constant output voltage (within certain limits) in spite of changing line voltage and/or load current.

■ **voltage source**

circuit or device that supplies voltage to a load.

■ **Voltage Standing Wave Ratio (VSWR)**

1. a comparison of the power transmitted to the power received back from the system with respect to a 50 Ohm line impedance. a VSWR of 1. 0 is a 100% efficient antenna, a VSWR of 1. 5 is 97%, a VSWR of 2.0 is 89%.

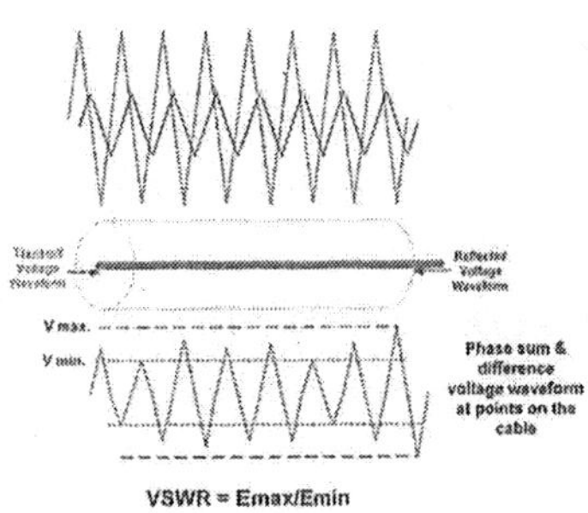

2. for a switching card, the loss due to the mismatch introduced into the signal by the card contacts and conductors. Expressed as a ratio of the highest voltage to the lowest voltage found in the signal.

■ **voltaic cell**

primary cell having two unlike electrodes immersed in a solution that chemically interacts to produce a voltage.

■ **volt-ampere**

unit of apparent power in an AC circuit containing capacitive or inductive reactance. Apparent power is the product of source voltage and current.

■ **voltmeter**

instrument used to measure difference in potential between two points.

■ **voltmeter**

a meter for reading voltage. It is one of the ranges in a multimeter.

■ **volume**

magnitude or power level of audio frequency. Measured in volume units (VU).

■ **VOX**

Voice Operated Switch.

■ **VXI**

VME bus extensions for instrumentation. an industry standard in test instrumentation.

■ **wafer**

a slice of silicon sliced from a 4, 5, 6 or 8 inch diameter silicon bar which is used as the foundation on which to build semiconductor products.

■ **wafer**

the thin (a few hundred microns) slice sawn from a cylindrical semiconductor crystal which serves as the substrate for the fabrication of multiple individual integrated circuits or chips.

■ **wafer probing**

an electrical test of devices on the wafer, using small probes to make contact with the metallised pads on the die.

■ **warm-up time**

the time required after power is applied to an instrument to achieve rated accuracy at referenced conditions.

■ **watt**

1. unit of electrical power required to do work at the rate of one joule per second. One watt of power is expended when one ampere of direct current flows

through a resistance of one ohm. In an AC circuit, true power is the product of effective volts and effective amperes, multiplied by the power factor.

2. a linear measure of optical power, usually expressed in milliwatts (mW), microwatts (*W) or nanowatts (nW).

■ wattage rating

maximum power a device can safely handle continously.

■ watt-hour

the unit of electric energy related to time. Thus, it represents 1 w of power dissipation for one hr.

■ wattmeter

instrument used to measure electric power in watts.

■ wave

1. an undulation or vibration, a form of movement by which all radiant energy of the electromagnetic spectrum is thought to travel.

2. electric, electromagnetic, acoustic, mechanical or other form whose physical activity rises and falls or advances and retreats periodically as it travels through some medium.

■ waveform

shape of a wave.

■ waveform (signal) generator

an oscillator that generates multiple output waveforms (sine, triangle, square waves) at the same frequency.

■ waveguide

rectangular or circular pipe used to guide electromagnetic waves at micro frequencies.

■ wavelength

1. the fundamental property of light—the length of the light wave, which determines its color. Common units of measurement (which is usually from crest to crest) are the micron, the nanometer, and (earlier) the angstrom.

2. a measure of the color of light, usually expressed in nanometers (nm) or microns (*m).

3. the physical distance that a radio wave travels during one complete cycle, the length is determined by the speed of light divided by the frequency of the wave.

4. The length of a wave based on the propagated wave distance in relation to one cycle. With a wave velocity of 300,000,000 meters per second (approximately 186,000 miles per second), 1 cycle 300,000,000 meters.

■ **wavelength division multiplexing (wdm)**
a technique of sending signals of several different wavelengths of light into the fibre simultaneously.

■ **weber**
(Wb) unit of magnetic flux. One weber is the amount of flux that when linked with a single turn of wire for an interval of one second will induce an electromotive force of one volt.

■ **wet cell**
secondary cell using a liquid as an electrolyte.

■ **wetting**
term used in soldering to describe the condition that occurs when the metals being soldered are hot enough to melt the solder so it flows over the surface.

■ **Wheatstone bridge**
four arm bridge circuit used to measure resistance, inductance or capacitance.

■ **wideband amplifier**
also called "broadband amplifier." Amplifier with a flat response over a wide range of frequencies.

■ **Wiedemann effect**
the mechanical torsion that occurs when anelectric current is passed along or through a long thin ferromagneticmaterial while it is subjected to an axial magnetic field.

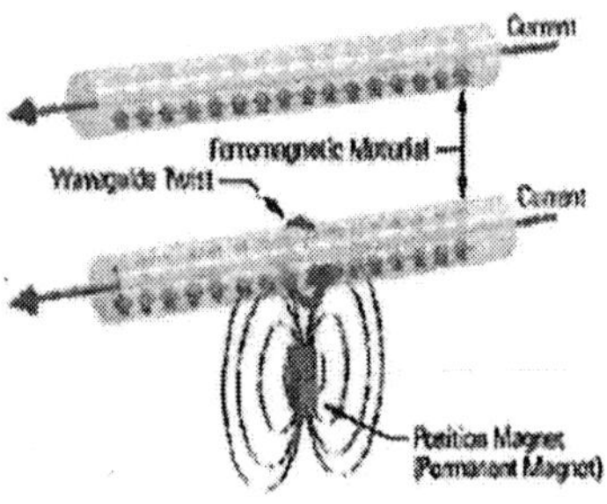

Wiedmann Effection a Ferromagnetic Conductor

■ **Wien-bridge oscillator**
oscillator that uses an RC low-pass filter and an RC high-pass filter to set the frequncy of oscillations.

■ **winding**
one or more turns of a conductor wound in the form of a coil.

■ **window**
a piece of glass with plane parallel sides which admits light into or through an optical system and excludes dirt and moisture.

■ **wire**
single solid or stranded group of conductors having a low resistance to current flow. Used to make connections between circuits or points in a circuit.

■ **wire gauge**
American Wire gauge (AWG) is a system of numerical designa-

tions of wire diameters.

wire wrapping
method of making a connection by wrapping wire around a rectangular pin.

wireless
term describing radio communication that requires no wired between two communicating points.

wirewound resistor
resistor in which the resistive element is a length of high resistance wire or ribbon usually nichrome wound onto an insulating form.

woofer
large loudspeaker designed primarily to reproduce low frequency audio signals.

word
the standard number of bits that a processor manipulates at one time. Microprocessors typically use 16-, or 32-bit words. (Or 2 bytes and 4 bytes respectively.).

work
work is done any time energy is transformed from one type to another. The amount of work done is dependent on the amount of energy transformed.

work function
the minimum energy required to remove an electron from the Fermi level of a material into field-free space. Work function is normally expressed in electron volts.

working margin
the difference (in db) between the power budget and the loss budget (i.e. The excess power margin).

working voltage
the maximum voltage at which a device will operate continuously with safety.

workstation
a somewhat generic term applied to any computer that has the performance and networking capability required to run the cad tools used in engineering design.

X
symbol for reactance.

X axis
horizontal axis.

Y
symbol for admittance.

Y axis
vertical axis.

Yagi antenna
an antenna system employing a basic antenna element as well as reflector and director rods. It is a highly directional antenna with

excellent signal pick-up characteristics.

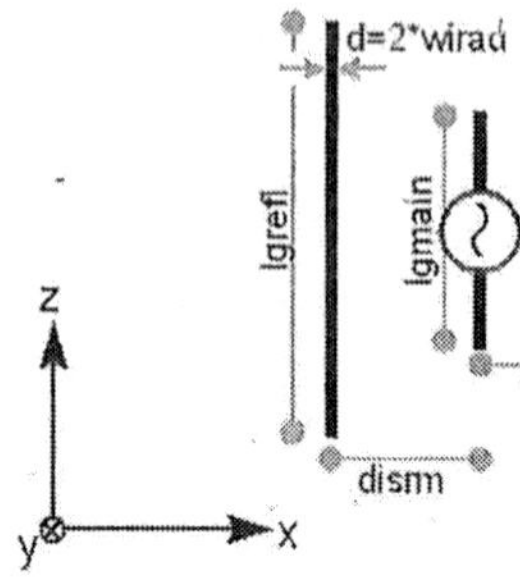

Construction of Yagi Antenna

■ yield

1. the percentage of good chips compared to the total chips produced per wafer. Yield generally increases as a particular process becomes tuned to the production of a particular chip but is limited by physical defects in the semiconductor crystal surface
2. the ratio of good units obtained divided by the total units produced. It is the percent of wafers, dice, or packaged units conforming to specifications.

■ yoke

in television, an inductor arrangement around the neck of the picture tube which provides electromagnetic deflection of the beam both vertically and horizont.

■ z axis

axis perpendicular to both X and Y axes.

■ zener diode

a diode characterised by having a breakdown region at a specific reverse-bias potential. This characteristic permits high current flow without diode damage. The breakdown region is known at the zener region and, despite the current increase, the voltage drop across the diode remains virtually the same as before the zener region is reached. Thus, the zener diode is useful in voltage-regulation applications.

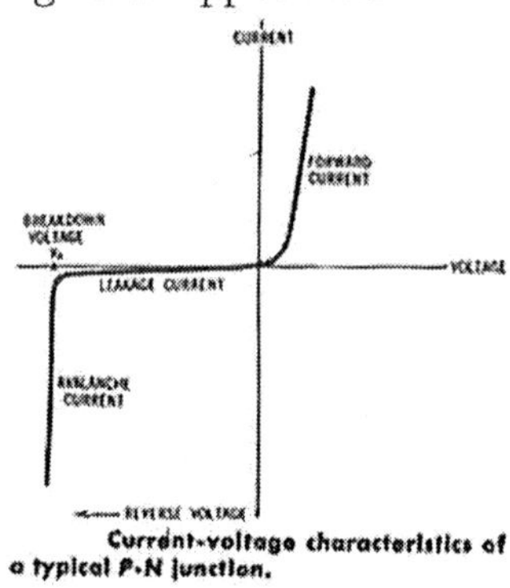

V-I Graph Showing Zener Region

■ Zerbst plot

a plot used to extract generation lifetime and surface generation velocity from c-t data. this plot is derived from d(1/c2) / dt vs. depletion width.

■ zero bias

the absence of a potential between the grid and cathode of a tube, or between the emitter—base or other electrodes of a transistor.

■ **zero offset**

the reading (desired or undesired) that occurs when the input terminals of a measuring instrument are shorted.

■ **zeroing**

calibrating a meter so that it shows a value of zero when zero is being measured.

NOTES

Other Books on

LOTUS ILLUSTRATED DICTIONARIES

1. First English Dictionary **(New)**
2. Astronomy
3. Agriculture
4. Anthropology
5. Archaeology
6. Architecture
7. Art
8. Banking Finance & Accounting
9. Bio - Chemistry
10. Business Administration
11. Bio - Technology
12. Biology
13. Botany
14. Culture
15. Chemical Engg.
16. Chemistry
17. Civil Engineering
18. Computer Science
19. Commerce
20. Cooking & Food
21. Ecology
22. Economics
23. Education
24. Electrical Engineering
25. Elcotronic & Telecommunication
26. Environmental Studies
27. Festival
28. Geography
29. Geology
30. Genetic Engineering

31. Health & Nutrition
32. History
33. Import and Export
34. Information System Management
35. Internet
36. IT
37. Inorganic Chemistry
38. Law
39. Library & Information Science
40. Literature
41. Management
42. Mathematics
43. Marketing & Sales
44. Mass Communication
45. Mechanical Engg.
46. Medical
47. Music
48. Organic Chemistry
49. Philosophy
50. Physical Education
51. Physics
52. Psychology
53. Science
54. Sex
55. Sociology
56. Sports
57. Textile
58. Zoology
59. Dictionary of Veterinary Sciences **(New)**
60. Dictionary of Synonyms and Antonyms **(New)**
61. Dictionary of Idioms and Phrases **(New)**